S357 S...

Space, Time and Cosmology

BLOCK 2

Electromagnetism and Einstein's special theory of relativity

Unit 4 *Electromagnetic fields*

Unit 5 *The need for special relativity*

Unit 6 *Some consequences of special relativity*

Unit 7 *Spacetime, momentum and energy*

Unit 8 *Consolidation and revision I*

S357 Course Team

Course Team Chair: Raymond Mackintosh

Raymond Mackintosh	Author	Michael Watkins	Course Manager
Russell Stannard	Author	David Tillotson	Course Editor
Bob Zimmer	IET	Peter Twomey	Course Editor
Tony Evans	Author, Assessor	Ian Thomas	Producer, BBC/OUPC
Leon Firth	Author	Liz Sugden	Assistant, BBC/OUPC
Bernard Schutz	Author	Tony Jolly	BBC/OUPC
John Charap	External Assessor	Steve Best	Graphic Artist
Gillian Stansfield	Assessor	Sarah Crompton	Graphic Design
Tom Smith	Reader	Hannah Brunt	Graphic Design
Alan Cooper	Reader	Alison Cadle	Manager OU TeX system

S357 is a revision of a course, S354, first presented in 1979. We should particularly like to acknowledge major contributions of the following original course team members:

John Bolton (OU), David Broadhurst (OU), Paul Clark (OU), Alan Cooper (OU), Tom Smith (OU), Russell Stannard (OU), Andrew Crilly (BBC), Al Saperstein, George Abell, Julian Schwinger.

List of Units

Block 1   Newtonian ideas of space and time
Unit 1   Space and time — the setting for the motion of a particle
Unit 2   Newtonian mechanics
Unit 3   Symmetries and conservation laws

Block 2   Electromagnetism and Einstein's special theory of relativity
Unit 4   Electromagnetic fields
Unit 5   The need for special relativity
Unit 6   Some consequences of special relativity
Unit 7   Spacetime, momentum and energy
Unit 8   Consolidation and revision I

Block 3   Gravitation, Einstein's general theory of relativity and black holes
Unit 9   First steps to a theory of gravitation
Unit 10 & 11   A metric theory of gravity and the field equations of general relativity
Unit 12   Black holes and other consequences of general relativity

Block 4   Cosmology and the early universe
Unit 13   The big bang
Unit 14   General relativity and cosmology
Unit 15   The evolution of the Universe
Unit 16   Consolidation and revision II

The Open University, Walton Hall, Milton Keynes, MK7 6AA.

First published 1997.

Edited, designed and typeset by the Open University using the Open University TeX System.

Printed in the United Kingdom by Henry Ling Ltd, at the Dorset Press, Dorchester, Dorset.

ISBN 0 7492 8159 6

This text forms part of an Open University Third Level Course. If you would like a copy of *Studying with The Open University*, please write to the Course Enquiries Data Service, PO Box 625, Dane Road, Milton Keynes, MK1 1TY. If you have not already enrolled on the Course and would like to buy this or other Open University material, please write to Open University Educational Enterprises Ltd, 12 Cofferidge Close, Stony Stratford, Milton Keynes, MK11 1BY, United Kingdom.

1.2

S357b2i1.2   *Title page photos of Newton (Block 1) and Maxwell (Block 2) are by courtesy of the Mansell Collection.*

Unit 4   Electromagnetic fields

Prepared by the Course Team

Contents

Aims

In this Unit, we intend to:

1 Show how the study of electromagnetic phenomena leads to a force law with a form that is incompatible with Newtonian mechanics — a conflict that points the way towards Einstein's extension of, and improvement on, Newton's ideas of space and time.

2 Clarify and emphasize the role of fields — a concept that assumes great importance later in the Course.

Objectives

When you have finished studying this Unit, you should be able to:

1 State Coulomb's law and, given the value of the constant ε_0, apply it to simple problems.

2 Give an operational definition of electric and magnetic field strengths.

3 Draw and interpret electric and magnetic field maps due to stationary charges and steady currents.

4 Distinguish between possible and impossible electric fields due to stationary charges.

5 Write a hundred words contrasting 'local' and 'action-at-a-distance' views of force; apply local field concepts to simple problems.

6 State the Lorentz force law and use it to draw vector diagrams of forces in some simple cases and solve simple problems.

7 Describe how changing electric and magnetic fields are related to one another, and how electromagnetic pulses propagate through space.

8 Explain how magnetic forces are inconsistent with the Galilean transformation of velocities and the form invariance of physical laws.

9 Explain how the speed of light, as derived from Maxwell's equations, poses problems for the Galilean transformation of velocities.

Study comment

In this Block, Unit 4 is rather shorter than the others, particularly Units 6 and 7. We suggest that you spend somewhat less than the usual study period on it in order to spend more time on the later Units.

1 Introduction to Block 2

Band 1 of AC2 introduces this Block.

For many years, it seemed that the theory of Newtonian mechanics was completely successful. It was based on plausible assumptions about space and time, and it led to results that seemed in complete agreement with experiment. The great French mathematician J. L. de Lagrange (1736–1813) summarized the feeling very well. 'There is', he said, 'only one Universe to discover, and Newton had both the genius and the good fortune to find it'.

However, in 1905, Einstein challenged Newtonian mechanics at its core. He claimed that it was constructed on inappropriate concepts of space and time. For example, in Unit 5 you will see that he rejected the assumptions that time intervals and distances are the same for all observers. In the new picture, the length of a moving rod is less than that of a stationary rod, and the time intervals between ticks on a moving clock are greater than those on a stationary one.

These extraordinary effects are very small at low speeds. An observer on the ground would find that the supersonic aircraft Concorde, flying at $2\,000$ kilometres per hour, 'shortens' by about 10^{-10} metres (the size of an atom) and its clocks 'lose' about 10^{-8} seconds every hour! But the effects *are* there, and become of major importance at high speeds.

Einstein's theory will be discussed in Units 5–7. But first, in Unit 4, we ask why Newton's 'common-sense' ideas were rejected and how Einstein could confidently predict such small discrepancies. We shall approach these questions by describing some discoveries that were made between the ages of Newton and Einstein, the most important of which were the laws of electricity and magnetism. These laws were unknown to Newton — they were accumulated, rather slowly, between 1785 and 1895 by a succession of physicists from Coulomb to Lorentz.

It will soon become clear that the laws of electromagnetism *cannot* be easily incorporated into Newtonian mechanics. Einstein was the first person to realize the full significance of this incompatibility. It provided him with his main motive in proposing special relativity, and it is not accidental that his historic 1905 paper was entitled, 'On the electrodynamics of moving bodies'. For these reasons, we shall devote the whole of Unit 4 to a review of the laws of electricity and magnetism. We shall then be ready, in Unit 5, to confront the problems that worried Einstein at the beginning of this century.

The conflict between Newtonian space and time and the laws of electricity and magnetism will become clear in Section 2 of Unit 5, which argues that simple experiments conducted with wires and batteries seriously threaten the Newtonian principle of the universality of distance. But what replaces Newton's view of the world?

Beginning with Section 3.3 of Unit 5, we adopt a rather axiomatic stance in order to explain Einstein's way of looking at things. We shall retain this approach for the rest of Block 2. The rewards are great, for we are able to present many explicit, quantitative examples of special relativity. These ought to help you towards an appreciation of Einstein's 1905 world-view; Einstein showed that time and space are not separated but are combined as four-dimensional spacetime.

2 Electric charge

We begin with a review of the concept of electric charge. This is an inherent property of some particles. It can be recognized because charged particles attract or repel one another with forces that are far stronger than those due to gravity. The charge carried by a particle is characterized by a single number, q, which may be positive or negative, and which depends only on the nature of the particle.

How are we to decide what number should be assigned to the charge carried by a chosen particle? Clearly we need an operational definition of charge.

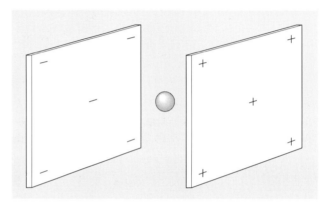

Figure 1   A charged particle held stationary between oppositely charged metal plates

One method might be to place the particle midway between two oppositely charged metal plates. *Both the metal plates and the particle are at rest in our reference frame.* We can then measure the force needed to prevent the particle moving towards either of the plates (see Figure 1). The ratio of the magnitudes of the charge on two particles can then be *defined* as the ratio of the forces needed to restrain them

$$\frac{q_1}{q_2} = \frac{|\mathbf{F}_1|}{|\mathbf{F}_2|}.$$

electric charge

The *magnitude* of the charge on any chosen particle is found by comparison with an arbitrarily chosen unit charge. The *sign* of the charge is defined from the direction of the force acting on it. The *convention* is to say that particles that accelerate in the same direction as electrons are *negatively* charged, while particles that accelerate in the opposite direction are *positively* charged.

This operational definition of charge allows us to check some of its properties. If we were to perform careful experiments based on this definition, we would discover the following laws:

conservation of charge

The conservation of charge: The total charge of a closed system does not depend on time. In particular, the charge on a particle is constant unless (e.g. as in β-decay) it changes into a different particle in which case the *total* charge is fixed.

additivity of charge

The additivity of charge: The charge of a composite particle is the sum of the charges of its component parts.

However, we cannot directly confirm that charge is an *invariant* under all circumstances, since the apparatus of Figure 1 has been explicitly assumed to be at rest. Instead, we *define* charge as an invariant quantity. So automatically, the charge of a particle does not depend on its motion. (It will become clear in Section 3.4 why this definition of charge invariance is useful.)

2.1   Coulomb's law of force between two stationary charges

The most important property of charges is the way they interact. The law of force between electric charges was discovered by means of systematic experiments performed by Augustin de Coulomb (1736–1806). He found that the force between two isolated stationary charges can always be described by the following formula, known today as Coulomb's law:

$$\mathbf{F}_{12} = \frac{1}{4\pi\varepsilon_0}\frac{q_1 q_2}{|\mathbf{x}_1 - \mathbf{x}_2|^2}\mathbf{e}_{12} \tag{1}$$

where $\mathbf{F}_{12}$ is the electric force on particle 1 due to particle 2, q_1 and q_2 are the charges assigned to the two particles, $|\mathbf{x}_1 - \mathbf{x}_2|$ is the distance between them, $\mathbf{e}_{12} = \dfrac{\mathbf{x}_1 - \mathbf{x}_2}{|\mathbf{x}_1 - \mathbf{x}_2|}$ is a vector of unit length pointing from particle 2 to particle 1 (see Figure 2), and $1/4\pi\varepsilon_0$ is a universal constant which, in SI units, has the value 8.988×10^9, but in all problems you should use the value 9×10^9. This corresponds to choosing a very large unit of charge — the coulomb, denoted C. Two charges each of $+1\,\mathrm{C}$ repel one another with a force of one newton when separated by about 100 kilometres!

ε_0 is pronounced 'epsilon nought' or 'epsilon zero'. The factor 4π is included in order to make certain other equations in electromagnetism simpler.

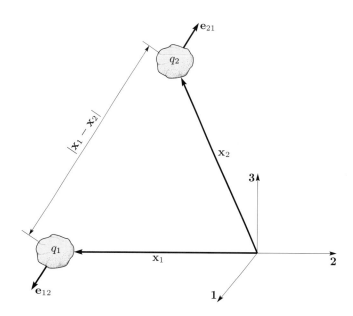

Figure 2   Charges q_1 and q_2 placed at positions $\mathbf{x}_1$ and $\mathbf{x}_2$

Coulomb's law bears a striking resemblance to Newton's law of gravitation. The magnitude of the force is proportional to the product of the two charges and decreases as the square of their separation. The direction of the force is always along the line joining the two particles (compare with Unit 2, Section 4.4). Both Coulomb's law and Newton's law of gravitation satisfy Newton's third law. They both give support to the principles of homogeneity and isotropy of space.

But there are also important differences:

(i)   Gravitational mass is always positive and gravitational forces are always attractive. But charged particles fall into *two* groups according to the *sign* of their charge. Members of the same group always repel one another, but members of opposite groups always attract one another.

(ii)   The electric force on charged elementary particles is *much* stronger than the force of gravity. For example, electrons have a charge-to-mass ratio of 1.8×10^{11} in SI units. For these particles, the ratio of electric to gravitational forces is

$$\frac{|\mathbf{F}|_{\text{Coulomb}}}{|\mathbf{F}|_{\text{Newton}}} = \frac{(4\pi\varepsilon_0)^{-1}}{G} \frac{q_1 q_2}{m_1 m_2} = \frac{9 \times 10^9}{6 \times 10^{-11}} \times (1.8 \times 10^{11})^2 \approx 10^{42}.$$

So why aren't all effects of gravity completely masked by those of electricity? The very strength of the electric force means that positive charges (nuclei) tend to draw negative charges (electrons) around them to form neutral atoms. On a macroscopic scale, the effects of these two different signs of charge cancel out. On the other hand, gravity is always attractive, so although it is intrinsically much weaker than an electric force, its effects are *additive* and so much more obvious to us on Earth. Everyday electrostatic effects correspond to *very* small imbalances between positive and negative charges.

2.2   The electric field produced by stationary charges

Both Coulomb's law and Newton's law of gravitation relate the forces between two particles *directly* to their positions. The underlying picture is that of so-called 'instantaneous action at a distance' which, from Newton onwards, has not appealed to physicists. However, it is possible to reformulate Coulomb's law in such a way that the idea of 'action at a distance' is avoided. This was first realized by Michael Faraday who, in the early nineteenth century, introduced the concept of an *electric field*. In this section, we explain what an electric field is, and discuss what form it takes around stationary charges.

The essential idea is that the space surrounding a charged body is, in some way, affected by the presence of the charge. The space is affected everywhere whether or not a second charged body is brought up to test it. But, by placing a test charge at a particular point, one can sense how the first charge influences that point in space.

In order to quantify the influence of an arbitrary arrangement of charges, we shall *define* the electric field at a given point to be the force per unit charge experienced by a small charge placed at that point. (We stipulate that the charge be small, just to be sure that it does not cause any redistribution in the charges whose field it is testing.) Our definition attributes to every position $\mathbf{x}$ in space an *electric field vector* $\mathbf{E}(\mathbf{x})$.

electric field vector, E

An electric field like $\mathbf{E}$ *will depend on position* $\mathbf{x}$. *This is indicated by writing* $\mathbf{E}(\mathbf{x})$. *However, this makes for clumsy looking equations, so the argument in parentheses is often omitted where there is no ambiguity. But remember that* $\mathbf{E}$ *and its components* E^1 *etc. are functions of the three position coordinates* x^1, x^2, *and* x^3, *i.e.* $\mathbf{x}$.

The direction and magnitude of this vector can be found by placing a stationary test charge q at the position $\mathbf{x}$ and measuring the force $\mathbf{F}$ it experiences. We define

$$\mathbf{E}(\mathbf{x}) = \frac{\mathbf{F}}{q} \qquad (2)$$

i.e. $(F^1, F^2, F^3) = (qE^1, qE^2, qE^3)$.

The electric field is a *local* property in the following sense: if we know $\mathbf{E}$ in some small neighbourhood, we know *without further enquiry* what will happen to a small test charge placed there. We have no need to know the particular arrangement of charges that has given rise to that field. The test charge responds to the local conditions in the space around it, and these are described by the field **E**. From Equation 2, one sees that the SI unit for **E** is newton per coulomb, $N\,C^{-1}$.

To predict the electric field produced by a given arrangement of charges, we go back to Coulomb's law. To take the simplest case, consider the electric field due to a single charge Q situated at the origin. According to Coulomb's law, if a test charge q is placed at the position **x**, it will experience a force

$$\mathbf{F} = \frac{1}{4\pi\varepsilon_0} \frac{qQ}{|\mathbf{x}|^2} \frac{\mathbf{x}}{|\mathbf{x}|}.$$

It follows that the electric field at the point **x** is

$$\mathbf{E}(\mathbf{x}) = \frac{\mathbf{F}}{q} = \frac{1}{4\pi\varepsilon_0} \frac{Q}{|\mathbf{x}|^2} \frac{\mathbf{x}}{|\mathbf{x}|}. \tag{3}$$

Because of the form of Coulomb's law, q has cancelled out from the right-hand side of Equation 3 so that $\mathbf{E}(\mathbf{x})$ is *independent* of the properties of the test charge. The field concept would be of little use if Coulomb's law involved, for example, $(q + Q)^2$ instead of qQ.

To be explicit, let us consider the field around an isolated positive charge. We can get a feel for the form of this field by sampling it at a few randomly chosen points (a, b, c, d, e and f in Figure 3) and drawing an arrow with its tail at each point in question.

The direction of the arrow indicates the direction of the field at that point (i.e. the direction of the force on a positive test charge). The length of the arrow is proportional to the magnitude of the field (i.e. to the magnitude of the force on the test charge).

This arrow map shows that the field is always directed away from the positive charge that creates it, and falls off with distance according to Coulomb's law.

electric field lines

An alternative way of visualizing an electric field is to use *field lines*. These are continuous lines which follow the direction of the electric field at each point along their length. For example, Figure 4(a) shows the field lines due to a single positive charge, and Figure 4(b) the field lines due to a single negative charge.

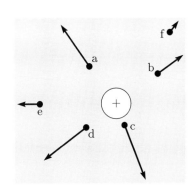

Figure 3   An arrow diagram for the field produced by a single positive charge. The arrows indicate by their length and direction the strength and direction of the field at the positions a, b, c, d, e and f.

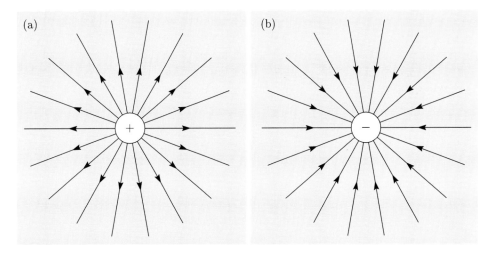

Figure 4   An alternative way, involving field lines, of representing the field produced by (a) a single isolated positive charge, and (b) a single isolated negative charge.

The electric field due to the simultaneous presence of more than one stationary charge can be found in exactly the same way — from the force exerted on a small test charge. If we know the distribution of the charges, we can use Coulomb's law and the *principle of superposition of forces* (Unit 2, Section 4.2) to predict what the field will be at any chosen point. For example, if the original charge distribution contains two charges Q_1 and Q_2 at positions $\mathbf{x}_1$ and $\mathbf{x}_2$, they will both contribute to the total force experienced by the test charge at position $\mathbf{x}$:

$$\mathbf{F} = \mathbf{F}_1 + \mathbf{F}_2$$

$$\therefore \quad \mathbf{E} = \frac{\mathbf{F}}{q} = \frac{\mathbf{F}_1}{q} + \frac{\mathbf{F}_2}{q} = \mathbf{E}_1 + \mathbf{E}_2$$

or, using Coulomb's law

the superposition of electric fields

$$\mathbf{E}(\mathbf{x}) = \frac{1}{4\pi\varepsilon_0} \left[\frac{Q_1}{|\mathbf{x} - \mathbf{x}_1|^2} \frac{\mathbf{x} - \mathbf{x}_1}{|\mathbf{x} - \mathbf{x}_1|} + \frac{Q_2}{|\mathbf{x} - \mathbf{x}_2|^2} \frac{\mathbf{x} - \mathbf{x}_2}{|\mathbf{x} - \mathbf{x}_2|} \right]. \tag{4}$$

Thus the total field at any given point is the vector sum of the individual fields due to the single charges. It is because of this that we say that the electric *field* obeys the principle of superposition.

We can use Equation 4 to draw electric field lines due to various arrangements of charge. In Figure 5, $Q_1 = -3$ and $Q_2 = +3$. In Figure 6, $Q_1 = -3.5$ and $Q_2 = +1.25$.

Several general points should be noticed about the field line diagrams of Figures 4–6:

(i) Field lines start on positive charges and terminate on negative ones (or go to infinity).

(ii) In our diagrams, we have used the *convention* that the number of lines starting or terminating on a given charge is proportional to the value of the charge.

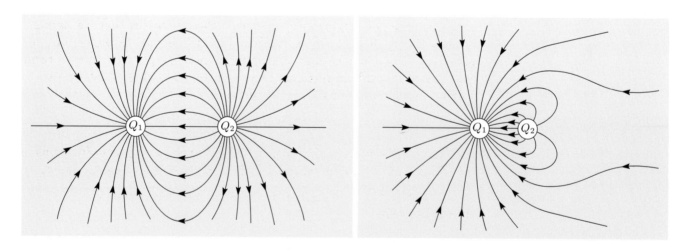

Figure 5 Field lines for $Q_1 = -3$ and $Q_2 = +3$

Figure 6 Field lines for $Q_1 = -3.5$ and $Q_2 = +1.25$

(iii) The field lines do not *directly* tell us the magnitude of the field, although we can see in a general way that the field lines tend to bunch together in regions of high field (near large charges) and to spread apart in regions of low field (away from the charges). One can go a little further in a case like Figure 4(a) where the lines have been drawn uniformly spaced on the surface of the charge. The magnitude of the field at a point is proportional to the density of the lines at that point. Figure 4 shows a projection onto a plane through a spherical field. It is clear that this field line picture is consistent with an inverse square law since the number of lines passing through the surface of a sphere of radius r, centred on the charge, is constant. Since the surface area of the sphere is $4\pi r^2$, it follows that the line *density*, i.e. the number of lines *per unit area*, varies as $1/r^2$, as required.

(iv)   The pattern of field lines has the same symmetry as the distribution of charges that produces it. Thus, the field lines in Figure 4 have the spherical symmetry of a point charge, while the field lines in Figures 5 and 6 are symmetric under any rotation about the line joining the two charges.

2.3   The idea of local field equations

The concept of an electric field allows us to think about electric forces in two separate stages:

(i)   The creation of the field by the charges.

(ii) The interaction of another charge with this field.

One can always calculate the field **E** due to a distribution of charge by applying Coulomb's law to each charge and then adding the fields, as the superposition principle allows. However, there is an alternative, equivalent, method. It is important since it is the form of the theory which can be generalized, leading eventually to Maxwell's comprehensive theory.

field equations

The alternative method of calculating **E** involves solving what are known as the *field equations* for **E**. In this course, we do not expect you to solve these equations.* We simply wish you to have some feeling for the phrase 'local field equations', and to appreciate how such equations can contain the same physics as Coulomb's action-at-a-distance law.

Study comment

One of the reasons for pausing to explain these points is that Einstein's general theory of relativity is also based on field equations which have a structure with some features similar to Maxwell's equations, 'source terms' for example. It should therefore help your understanding of Block 3 to follow through our general discussion here.

The essential idea we wish to convey is that there are some quite stringent conditions on the way in which electric fields vary from point to point.

First, clearly from Figures 4–6 the electric field does not vary discontinuously.

Second, we may use the law of conservation of energy to rule out certain field configurations. To see this, consider the arrangement of test charges shown in Figure 7. If this arrangement of charges is placed in any of the fields shown in Figure 8, the forces acting on it will cause it to spin. If we keep all the charges responsible for the field stationary, it is difficult to see where this kinetic energy could come from. We must conclude, therefore, that a constant electric field, created by stationary charges, *cannot* vary from point to point in the manner of Figure 8(a)–(d).

*A full account is given in the current, 1997, OU course SMT356 *Electromagnetism*.

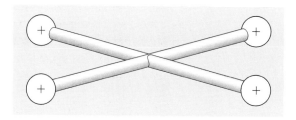

Figure 7   An arrangement of four equal test charges connected by rigid rods

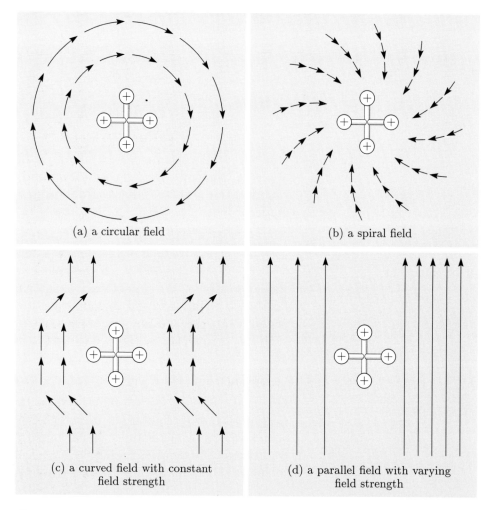

(a) a circular field

(b) a spiral field

(c) a curved field with constant field strength

(d) a parallel field with varying field strength

Figure 8   Field line diagrams of some impossible electric fields. In (a) to (c), the arrows indicate direction only. In (d), the field strength is greatest where the spacing is least.

So far, we have seen that the requirements of continuity and conservation of energy impose restrictions on possible electric fields. But this cannot be the whole story — for instance, no explicit use has been made of the fact that the force between two charges decreases as the *square* of their distance apart. We shall now see how Coulomb's law can be incorporated into the local field equations.

In order to do this, consider Figure 9. This shows, in cross-section, a charge, with two imaginary spheres constructed around it. We notice that, while the magnitude of the electric field at the surface of each sphere *decreases* as the square of its radius (Coulomb's law), the surface area of each sphere *increases* as the square of its radius. Thus, the quantity

$$\Phi = |\mathbf{E}| \times \text{(area of sphere)} = \left(\frac{1}{4\pi\varepsilon_0} \frac{Q}{r^2} \right) 4\pi r^2 = \frac{Q}{\varepsilon_0}$$

has the same value for *both* spheres centred on this charge. This is actually part of a far more general statement: if we place a sphere among *any* arrangement of charges, some within it and some outside it (as in Figure 10), the *average radial component*, $\overline{E}_r$, of the electric field over the surface of the sphere always obeys the equation:

$$\Phi = (\overline{E}_r) \times (\text{area of sphere}) \ = \frac{(\text{total charge contained in sphere})}{\varepsilon_0}. \quad (5)$$

(The quantity Φ is frequently called the total *electric flux* over the sphere.)

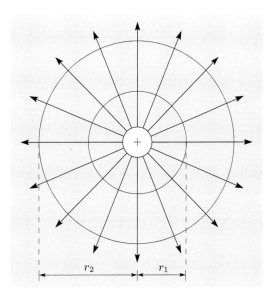

Figure 9 Cross-section of field lines from a single positive charge passing through two spheres of radius r_1 and r_2 centred on the charge

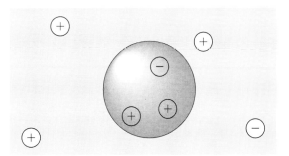

Figure 10 A sphere placed among an arbitrary arrangement of charges

Study comment

The justification of this result is sketched below. If you are prepared to take it on trust, you could omit the following 'proof'. You are certainly not expected to reproduce it.

Non-assessable ▼
optional text

Because of the superposition of electric fields, we really only have two cases to discuss:

1   A single charge somewhere outside a sphere, as in Figure 11(a);

2   A single charge somewhere inside a sphere, as in Figure 11(b).

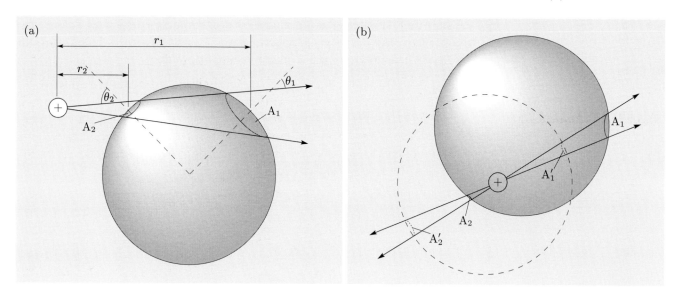

Figure 11   An electric charge situated (a) somewhere outside the sphere, and (b) somewhere inside the sphere

We shall first show that the average radial field due to the point charge in Figure 11(a) is zero. To do this, draw a cone from the charge through the sphere and consider the two small areas A_1 and A_2, at distances r_1 and r_2 from the charge. At A_1, the field lines flow out of the sphere, so in this region the *radial* component of **E** is positive. On the other hand at A_2, the field lines flow inwards, so here the radial component of **E** is negative.

The magnitude of the field at A_1 is weaker than the magnitude of the field at A_2 by a factor of $(r_2/r_1)^2$.

Hence the radial component of the field at A_1 is weaker than the radial component of the field at A_2 by a factor

$$\left(\frac{r_2}{r_1}\right)^2 \frac{\cos\theta_1}{\cos\theta_2}.$$

However, this is exactly balanced by the fact that the area of A_1 is greater than the area of A_2 by a factor

$$\left(\frac{r_1}{r_2}\right)^2 \frac{\cos\theta_2}{\cos\theta_1}.$$

This argument can be repeated for all cones from the charge which intersect the sphere; we conclude that $\overline{E}_r = 0$. This is a special case of Equation 5 since the total charge contained by the sphere is zero.

A similar argument applies to Figure 11(b) in which a positive charge is at an arbitrary point inside the sphere. The electric flux Φ does not depend on the position of the charge within the sphere. To prove this, consider a second sphere, shown in dashed lines in Figure 11(b), which has the same radius as the first.

This time, A_1 and A_1' contribute equal amounts to the average radial fields on their respective spheres — and so do A_2 and A_2'. Thus, the value of $\overline{E}_r$ is the same for *any* two spheres of the same radius. But there is one case in which it is easy to calculate Φ. When the charge is at the centre of the sphere (as in Figure 9) we know that

$$\Phi = \overline{E}_r \times (\text{area of sphere}) = Q/\varepsilon_0$$

but this has now been established *no matter where* the charge is placed within the sphere.

End of optional text   ▲

From our point of view, the importance of Equation 5 is that it re-expresses Coulomb's law in a local way. We can now take any point in a field, draw a small sphere around it, and calculate the value of Φ from the total charge enclosed by the sphere. *According to Equation 5, Φ is entirely determined by the amount of charge within the sphere; it is independent of the presence of charges outside the sphere.*

three local conditions for E   So we now have *three* conditions that must be satisfied by the electric field $\mathbf{E}(\mathbf{x})$ due to a stationary arrangement of charges. In rough terms, we may say that, in the *empty space* between the charges, $\mathbf{E}(\mathbf{x})$ should vary continuously; it should not 'circulate' in any particular sense (as in Figure 8a), nor should it 'diverge' (as in Figure 12).

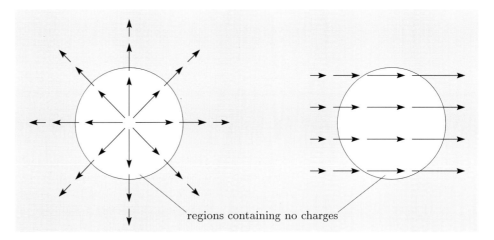

regions containing no charges

Figure 12   Some impossible diverging electric fields

Each of these conditions tells us about the rate of change of the field from one point to the next.

Figure 13 is meant to convey this idea: if one knows the field within the region R_1, one can calculate its value within a neighbouring region R_2, and so on, across all space.

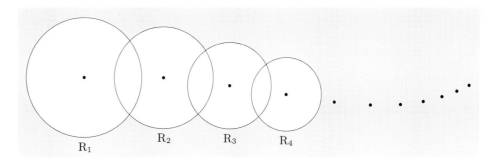

R_1 R_2 R_3 R_4

Figure 13   Calculating fields within neighbouring regions, successively across all space

But what determines the field in R_1 in the first place? Of course, if R_1 were centred on a completely isolated charge, this question would be easy to answer: $\mathbf{E}$ would be simply determined by Equation 5. But what happens in more general situations (such as Figures 5 or 6) in which more than one charge is present? Fortunately, this introduces no additional problems, since the principle of superposition tells us that the total field in any local region due to a number of charges is the sum of the contributions of each charge. We therefore find $\mathbf{E}$ for each charge and then add the contributions.

We conclude that it *is* possible to specify the electric field in a purely local way: each charge influences the field in its *immediate* environment, and this then determines the field a little further on, and so on. It is in this sense that the words 'action at a distance' can be eliminated from the vocabulary of fields.

Objectives 1 and 2   SAQ 1   Find, to an order of magnitude, the electric force between two protons in a helium nucleus. A proton has a charge of 1.6×10^{-19} C and the diameter of a helium nucleus is about 10^{-15} m. (Use $(1/4\pi\varepsilon_0) = 9 \times 10^9$ in SI units.)

What is the strength of the field produced by each proton at the position of the other?

Objective 4   SAQ 2   In Figure 8, various impossible fields were shown which varied in direction and magnitude. Explain carefully why the field shown in Figure 5 would not violate energy conservation even though it too involves fields that vary in direction and magnitude. (Consider placing the arrangement of charges of Figure 7 at various positions in Figure 5 and the forces that would be experienced on each of the four charges.)

Objective 5   SAQ 3   In Unit 2, Section 4.4, we mentioned that the *gravitational* force experienced by an apple above the Earth's surface would be 'unchanged if the Earth were shrunk to a point, with all its mass concentrated at its centre'. By *analogy* with methods developed in the section you have just read, can you suggest an argument to establish this result?

Objective 5   SAQ 4   Figure 14 shows a long wire, which carries a uniform distribution of positive charge and which lies along the diameters of two imaginary concentric spheres.

Compare the average radial electric field of each of these spheres.

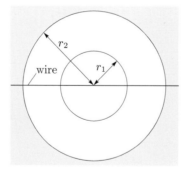

Figure 14 Two spheres, radii r_1 and r_2, with common centre on a horizontal wire carrying a uniform charge density

2.4   The field equations of electrostatics

In Section 2.3, we stated that the content of Coulomb's law could be re-expressed in terms of field equations for the electrostatic field **E**. These equations embody in mathematical form all the behaviour of **E**, including the prohibition on fields such as those shown in Figures 8 and 12.

Study comment

This course has been devised to avoid the use of partial differentials in all examinable material. The field equations cannot be written without partial differentials. You will therefore not be expected to reproduce or perform manipulations on the equations in this section. However, we feel that seeing the general structure will be helpful to you when we present the somewhat similar field equation for general relativity. If you are unfamiliar with partial differential notation, we briefly define it here.

Suppose U depends on a number of variables, perhaps on x^1, x^2, x^3, t so that it would be written in full as $U(x^1, x^2, x^3, t)$. Then $\dfrac{\partial U}{\partial x^1}$ represents the rate at which U changes as x^1 changes with (in this case) x^2, x^3 and t held fixed. It is called the 'partial differential of U with respect to x^1.'

The electrostatic field at point $\mathbf{x}$, $\mathbf{E}(\mathbf{x}) = (E^1, E^2, E^3)$, due to some arbitrary distribution of charge can be written as a solution to certain differential equations called the electrostatic field equations, which we now write down.

Let us say that electric charge is distributed according to a *charge density* $\rho(\mathbf{x})$. This is the charge within a unit volume at point $\mathbf{x} = (x^1, x^2, x^3)$; in other words, the charge within some infinitesimal volume $dx^1\, dx^2\, dx^3$ at $\mathbf{x}$ is $\rho(\mathbf{x})\, dx^1\, dx^2\, dx^3$. Then the electric field $\mathbf{E}$ at *any* point $\mathbf{x}$ is determined by two equations which we shall first write in conventional shorthand form, and then explain the notation. The first equations is:

$$\operatorname{curl} \mathbf{E} = \mathbf{0}. \tag{6}$$

By definition, $\operatorname{curl} \mathbf{E}$ is a vector field whose x^1 component is

$$(\operatorname{curl} \mathbf{E})^1 = \frac{\partial E^3}{\partial x^2} - \frac{\partial E^2}{\partial x^3}.$$

The x^2 and x^3 components of $\operatorname{curl} \mathbf{E}$ are found by cycling $1 \to 2 \to 3 \to 1$ so that:

$$(\operatorname{curl} \mathbf{E})^2 = \frac{\partial E^1}{\partial x^3} - \frac{\partial E^3}{\partial x^1}, \qquad (\operatorname{curl} \mathbf{E})^3 = \frac{\partial E^2}{\partial x^1} - \frac{\partial E^1}{\partial x^2}.$$

We can regard Equation 6 as consisting of three equations in each of which a different one of the three components of $\operatorname{curl} \mathbf{E}$ is set equal to zero.

The second equation is:

$$\operatorname{div} \mathbf{E} = \frac{\rho}{\varepsilon_0} \tag{7}$$

where $\operatorname{div} \mathbf{E}$ is defined as the particular sum of differentials:

$$\operatorname{div} \mathbf{E} = \frac{\partial E^1}{\partial x^1} + \frac{\partial E^2}{\partial x^2} + \frac{\partial E^3}{\partial x^3}. \tag{8}$$

From the definition, Equation 8, we see that $\operatorname{div} \mathbf{E}$, the 'divergence of $\mathbf{E}$', is a scalar field derived from the vector field $\mathbf{E}$.

Although as emphasized in the study comment, you will *not* be asked to reproduce Equation 6 and Equation 7, you should note their structure. They are differential equations involving the three components E^1, E^2, E^3 of $\mathbf{E}(\mathbf{x})$. They express the manner in which they vary together from point to point in space under the influence of a 'source'. The right-hand side of each equation is a 'source term.' At present, the only source of the field **E** is the charge density ρ; later, we shall find a new source, and that will appear on the right of the 'curl' equation. At present, the 'curl' equation, Equation 6, ensures that the field **E** does not take on impossible configurations like those in Figure 8. Equation 7 rules out the other class of impossible field configurations shown in Figure 12.

Equation 6 and Equation 7 together are an alternative way of writing Coulomb's law for the field **E**. At a point entirely outside the region where there is charge, $\rho = 0$; consequently div **E** $= 0$ everywhere outside a distribution of charge. Naturally, for a single charge, we can get back to Coulomb's law and, outside a spherical distribution of charge, we again just get the inverse square law (recall that as noted above, the field outside such a distribution is the same as if the charge were concentrated at a point).

In Units 10 & 11, you will see that Einstein's field equations for general relativity also take the form of differential equations, but in that case, we are dealing with a gravitational distortion of spacetime rather than an electrostatic field, and the source term is, not surprisingly, (essentially) the *mass* density, rather than the electric charge density.

3 Charges in motion

Up to this point, we have discussed interactions between stationary charges only. We now turn to consider more general situations in which the charges can be moving, both relative to one another and relative to the observer. We are particularly concerned to see if such motion makes any difference to the size and direction of the forces acting between particles. Before discussing the experimental evidence, we shall first remind you of some expectations raised by the Newtonian mechanics of Block 1.

3.1 Newtonian ideas about velocity-dependent forces

One of the basic ideas in Newtonian mechanics is that bodies can respond to forces which act instantaneously across empty space. For example, Newton's law of gravitation (Unit 2, Section 4.4) states that the gravitational attraction between two particles at a given instant is related to the positions of the particles *at that precise instant*:

$$|\mathbf{F}_{12}(t)| = \frac{Gm_1 m_2}{|\mathbf{x}_1(t) - \mathbf{x}_2(t)|^2} = m_1 |\mathbf{a}_1(t)|.$$

As we discussed in Unit 2, this law contains one very strange feature: it seems to imply that the acceleration of a particle is determined entirely by the instantaneous positions of other particles. A particle in one place seems to know without delay the position of another particle in another place.

It is natural to feel uneasy about such a strange law. In fact, Newton himself regarded the law of gravitation as no more than an empirical description:

From *The mathematical principles of natural philosophy (Principia)* (1729), translated by Andrew Motte.

> Hitherto I have not been able to discover the cause of these properties of gravity from phenomena, and I frame no hypotheses.

But most physicists did not allow themselves to be swayed by vague feelings of unease. It could not be denied that Newton's laws worked with admirable precision. They accounted for a vast range of quantitative phenomena — from planetary motion to the path of a cannonball. This unparalleled success persuaded many people that it would eventually become possible to explain all phenomena, including those of electricity and magnetism, in terms of fundamental forces acting instantaneously through empty space.

Let us therefore suppose that, at any given time, the force between two charged particles depends only on their instantaneous properties — their positions, velocities, charges, etc. — at the instant concerned. However, for the sake of generality, we shall now raise the possibility that *moving* charges do not interact according to Coulomb's law, but that electrical forces depend on both the positions *and the velocities* of the particles, so that Coulomb's law is only valid for *stationary* particles. But, in doing so, we should also remember to take account of the principle of relativity. One way of applying this principle was explained in Section 2.1 of Unit 3: all inertial observers are equivalent, so one would expect them to discover laws of the same form. In particular, fundamental laws should be form invariant under uniform linear boosts.

This seems to rule out proposed laws of the type:

$$|\mathbf{F}| = k|\mathbf{v}_1|\,|\mathbf{v}_2|.$$

If such a law were to hold, then two inertial observers related by a uniform linear boost:

$$\mathbf{v}_1' = \mathbf{v}_1 - \mathbf{u} \qquad \text{and} \qquad \mathbf{v}_2' = \mathbf{v}_2 - \mathbf{u}$$

would find that

$$|\mathbf{F}'| = k|\mathbf{v}_1'|\,|\mathbf{v}_2'| \quad \neq \quad k|\mathbf{v}_1|\,|\mathbf{v}_2| = |\mathbf{F}|.$$

This is inconsistent with Newtonian concepts of space and time because, in Newtonian mechanics, all inertial observers must agree about the magnitudes of accelerations, and hence about the magnitudes of forces.

Indeed, it is easy to see that, if forces depend on velocities at all, they should only do so via the *relative* velocity, this being unchanged by uniform linear boosts:

$$\mathbf{v}_1' - \mathbf{v}_2' = (\mathbf{v}_1 - \mathbf{u}) - (\mathbf{v}_2 - \mathbf{u}) = \mathbf{v}_1 - \mathbf{v}_2.$$

In the last section of Unit 3, we arrived at an even stronger statement: provided the energy function of a system splits into separate kinetic and potential energies, as in Equation 15 of Unit 3, the forces should not depend on velocity at all. Newton's law of gravitation provides one example of this, and one might expect that electrical effects should be described by a similar law — that Coulomb's law should apply no matter how the observer or the particles are moving. But, of course, such questions can only be answered by careful experiments.

From the point of view of Newtonian mechanics, the experimental results are quite shocking. Coulomb's law does not apply to moving particles — the force between two charges *does* depend on their velocities, *and* it does so via combinations like $|\mathbf{v}_1|\,|\mathbf{v}_2|$ rather than $(\mathbf{v}_1 - \mathbf{v}_2)$!

These facts are of the greatest interest to us since they cannot be reasonably explained within Newtonian ideas about space and time. Nowadays, we understand them in terms of Einstein's theory of relativity. Like the effects on Concorde, mentioned in the Introduction, the corrections to Coulomb's law are often negligible — for example if the two particles move at a million kilometres per hour, Coulomb's law will be wrong by about one part in a million. Yet all you need to demonstrate the existence of such effects are a battery and some lengths of wire!

3.2   Velocity-dependent forces between charges

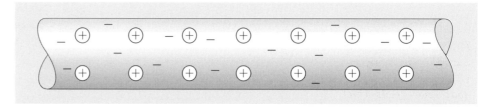

Figure 15   The structure of a metal wire, showing fixed positive ions and freely moving negatively charged electrons

Figure 15 is a schematic drawing of the structure of a wire, more or less as we understand it today. We regard it as being composed of electrons, which are not bound to particular atoms but are free to move around, together with a fairly rigid arrangement of positive ions. The total charge on the electrons and ions, of course, adds up to zero, so that overall the wire is neutral. A positive test charge placed near the wire will be neither attracted nor repelled, the attraction due to the electrons being exactly counterbalanced by the repulsion due to the positive ions. In other words, the electric field due to the wire is zero.

Now suppose we connect the wire to the two terminals of a battery (Figure 16). This will cause the electrons to flow along the wire, while the positive ions will remain practically fixed.

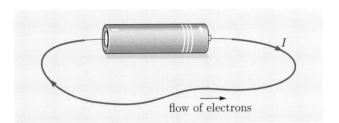

Figure 16   A wire connected to the terminals of a battery

electric current

So, we have a flow of charge along the wire, due to the movement of the electrons. The current is defined to be the rate of flow of charge. It has a magnitude given by the expression

$$I = nev, \tag{9}$$

where n is the number of electrons per unit length of wire, e is the magnitude of their charge, and v is the average speed at which they flow along the wire. But the *direction* of the current is *opposite* to that of the electrons' motion because, by *convention*, electrons carry a negative charge, $-e$ (see Section 2). This convention has its origins long before the electron was discovered.

The microscopic details of the conduction may be quite complicated — all the electrons interact with each other and collide with the positive ions. But, for our purposes, the simple picture of a steady flow of electrons through an immobile cage of ions is adequate.

We now ask whether the electric forces caused by the moving electrons still cancel those due to the static positive ions. Is the electric field due to the wire still zero? There is only one way of settling this issue — we must place a stationary test charge close to the wire, and see if it experiences a net force.

For the moment, we restrict ourselves to steady currents. Changing currents give rise to additional effects which will be discussed in Section 4.1.

If we were to perform this experiment, we would find that the stationary test charge still experiences no force due to the wire. In other words, the electric field caused by the wire is zero *whether or not a steady current is flowing.* Of course, from the viewpoint of Block 1, this is not very surprising — it is entirely consistent with the presence of instantaneously transmitted forces which are independent of velocity.

But suppose we perform a slightly different experiment. Instead of keeping the test charge fixed at a single point in space, we now allow it to move. For example, the test charge might be an electron which moves through the tube of a television set until it strikes the screen. In this case, we find a quite extraordinary effect — *the motion of the test charge is affected by the presence of currents in nearby wires.*

Figure 17 shows three different examples in which an electron is initially moving parallel to a wire. If no current flows through the wire (Figure 17a), no forces will act on the electron and it will move in a straight line. However, if a current is flowing through the wire, the electron will move along a different path. Depending on the direction of the current, it is either attracted or repelled by the wire (see Figures 17b and 17c).

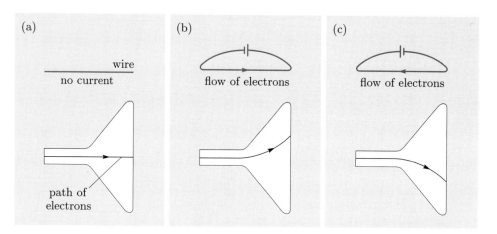

Figure 17   Movement of a test charge (an electron) in the vicinity of a wire.
(a) No force experienced by the test charge due to the wire.
(b) The test charge is attracted to the wire.
(c) The test charge is repelled by the wire.

magnetism

magnetic field

We have already seen that the wire produces no electric field, so another kind of force must be involved. This is the force of *magnetism*, and we say that the moving charge is affected by the *magnetic field* created by the currents in the wire.

Magnetic objects, such as compass needles and bar magnets, are also affected by currents in wires. This fact has been known since about 1820, when Oersted saw a compass needle deflect as he brought it near a current-carrying wire. In fact, a coil of wire behaves exactly like a bar magnet, and this has been used to construct commercial electromagnets. In this Unit, we shall concentrate on the magnetic effects created and detected by moving charges. We have no need to explain the origin of the magnetism of a piece of iron, which turns out to be a very complicated matter which can only be explained using quantum mechanics.

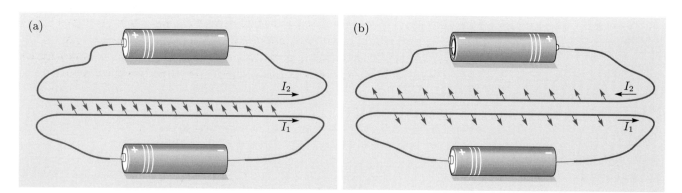

Figure 18   Parallel wires carrying currents are (a) pulled together when the currents are in the same direction, and (b) pushed apart when the currents are in opposite directions.
(For an easily detectable effect, quite large currents are required.)

The effect shown in Figure 17 can be observed with even simpler equipment. For example, we can place two wires side by side and connect them to batteries, as in Figure 18(a) and (b). In this case, the electrons in each wire behave as moving test charges which are attracted or repelled by the current in the opposite wire. Depending on the direction of the currents, the wires are either pushed apart or pulled together. In either case, the magnitude of the force per unit length of wire is given by the expression

$$\frac{|\mathbf{F}|}{l} = 2 \left(\frac{\mu_0}{4\pi} \right) \frac{I_1 I_2}{d}, \tag{10}$$

where $\mathbf{F}$ is the total force, l the length of the wire, I_1 and I_2 are the currents in the two wires, d is the distance between them, and $\mu_0/4\pi$ is a universal constant that characterizes the strength of magnetic forces (in SI units, it is precisely 10^{-7}).

These effects are really very strange. By combining Equations 9 and 10, it is clear that the magnetic force between two wires depends on the *product* $|\mathbf{v}_1||\mathbf{v}_2|$ of the average speeds of the electrons in the two wires. Not only does this imply velocity-dependent forces, but this dependence cannot be expressed in terms of the relative velocity $(\mathbf{v}_1 - \mathbf{v}_2)$. For example, the relative velocity of the electrons can be zero in two quite different cases:

(a) for two wires carrying no currents,

(b) for two identical wires carrying identical currents.

In the first case there is no force between the wires, but in the second case the wires attract one another!

It is clearly going to be very difficult to reconcile these effects with Newtonian views about space and time. However, before turning to discuss this conflict in Unit 5, we must give a more quantitative description of the fundamental laws that control magnetic phenomena.

3.3   The magnetic field

In Section 2.2, we saw that the force acting on a stationary test charge q could be expressed in terms of an electric field, $\mathbf{E}$. The force is $\mathbf{F} = q\mathbf{E}$, where $\mathbf{E}$ is the electric field in the immediate vicinity of the charge.

In this section, we shall try to use the same language of fields to express the effect of a current on a moving charge. Consider, for example, the coil and battery shown in Figure 19. We can say that the current in the wire affects the properties of space around it by creating a magnetic field. A moving test charge then responds to the local magnetic field in its immediate vicinity. By analogy with the electric field, we can quantify the magnetic field in terms of the *force* experienced by the test charge.

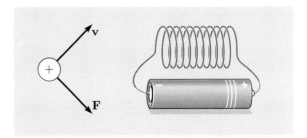

Figure 19   An arrangement of coil and battery producing a magnetic force on a test charge

In practice, the relationship between magnetic forces and magnetic fields is more complicated than in the electric case — the magnetic force depends on both the field *and* the velocity of the test charge. Nevertheless, a series of experiments, with different test charges moving with different velocities, would show that the magnetic force can always be written in the form:

$$\mathbf{F} = q(\mathbf{v} \times \mathbf{B}) = q(v^2 B^3 - v^3 B^2, v^3 B^1 - v^1 B^3, v^1 B^2 - v^2 B^1) \qquad (11)$$

where q is the charge of the test particle, $\mathbf{v} = (v^1, v^2, v^3)$ is its velocity, and $\mathbf{B} = (B^1, B^2, B^3)$ is a vector which varies from place to place, and with the current passing through the wires, but which is the same for all test particles. In Equation 11, we have written out the the vector product in component form.

The vector $\mathbf{B}$ characterizes the influence of the currents on the space around them. It provides us with a suitable way of quantifying the magnetic field, and we shall refer to it as the *magnetic field vector*. If we know the value of $\mathbf{B}$ at a particular point in space, we can use Equation 11 to calculate the magnetic force on *any* charged particle moving with *any* velocity through that point.

magnetic field vector, B

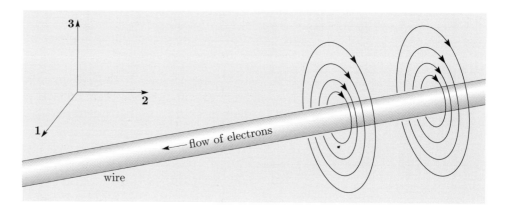

Figure 20 Field lines showing the direction of the magnetic field around a current-carrying wire

As an example of a magnetic field, consider Figure 20. This shows some lines of the magnetic field near a current-carrying wire. The magnetic field lines are circles surrounding the wire. These lines show the direction of the magnetic field (*not* of the magnetic force). The magnitude of the field is inversely proportional to the radial distance, r, from the wire. In fact

$$|\mathbf{B}| = 2\left(\frac{\mu_0}{4\pi}\right)\frac{I}{r}, \qquad (12)$$

where I is the magnitude of the current. The SI unit of $\mathbf{B}$ is the tesla (symbol T).

The effects of this field can be detected by moving a test charge in the region close to the wire. If the test charge moves parallel to the wire, it will experience a force at right-angles to both $\mathbf{v}$ and $\mathbf{B}$. That is, it will either be attracted to or repelled by the wire, as in Figures 17 (b) and (c). But if the test charge moves as in Figure 21, with $\mathbf{v}$ and $\mathbf{B}$ parallel to each other, the charge will experience no magnetic forces. Both the direction and the magnitude of the force depend on the velocity of the test charge.

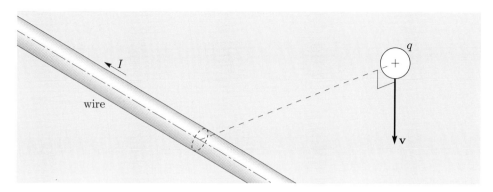

Figure 21 A test charge moving (momentarily) in the direction of the magnetic field about a current-carrying wire

Objective 3   SAQ 5   Show that Equation 10, for the force between two currents, can be interpreted in terms of the magnetic field illustrated in Figure 20. [*Hint:* Derive Equation 10 from Equation 12 using Equation 9 and Equation 11.]

Objective 3   SAQ 6   (a) Consider a steady current of electrons along a straight wire. What experimental evidence supports the claim that the *electric* field due to the stream of moving electrons is independent of their velocity?

(b) Show that the *magnetic* field due to a steady current of electrons spaced at n per metre in a wire can be written in the form

$$\mathbf{B}(\mathbf{x}) = \varepsilon_0\mu_0(\mathbf{v} \times \mathbf{E}_e(\mathbf{x}))$$

where $\mathbf{v}$ is the average velocity of the electrons, and $\mathbf{E}_e$ is the electric field due to the electrons at rest whose magnitude is given by the formula $|\mathbf{E}_e| = 2ne/(4\pi\varepsilon_0 r)$. [*Hint:* First calculate $|\mathbf{B}|$ at a distance r from the wire, and then consider the direction of $\mathbf{B}$.]

Objective 7   SAQ 7   The electric field at a given point is zero. The magnetic field is given by $\mathbf{B} = (2, 3, 5)$. What is the force on a particle of unit charge which moves with velocity $\mathbf{v} = (1, 2, 3)$ through this point? What is the magnitude of this force? What is the maximum magnitude of the force that can be exerted by this field on the same particle moving with the same speed? (All quantities are measured in SI units.)

3.4 The Lorentz force law

So far, we have considered the *separate* effects of electric and magnetic fields. In Section 2.2, we saw that a stationary arrangement of charges produces an electric field $\mathbf{E}$. This field determines the electric force on any stationary test charge. We have just seen that a steady flow of currents through neutral wires gives rise to a magnetic field $\mathbf{B}$, and this tells us the magnetic force on any moving test charge. But, in general, both electric and magnetic fields are present, and we must take account of both of them. In this case, the force on a test particle of charge q is given by the *sum* of the electric and magnetic forces:

$$\mathbf{F} = q\mathbf{E} + q\mathbf{v} \times \mathbf{B}$$
$$= q(\mathbf{E} + \mathbf{v} \times \mathbf{B}). \tag{13}$$

Lorentz force law

Figure 22   Hendrik Antoon Lorentz, 1853–1928. Einstein said of him: 'He meant more to me than anyone else I have met in my lifetime.'

*Of course, **F** and its components F^1 etc., like **E** and **B** and their components, are all functions of position, $F^1(\mathbf{x})$, etc. But in keeping with usual practice, and in defence of the readability of equations, we often omit them. See marginal note on p. 6.*

This equation summarizes much of our discussion so far, and nowadays is known as the *Lorentz force law*, see Figure 22. Note that the sign of the charge, q, determines the direction of the force.

It is important to realize that Lorentz's law has two distinct aspects — it is both a *definition* and a *physical law*.

(i)   *As a definition*

Lorentz's law contains general operational definitions of **E** and **B**. In order to see this, remember that the charge q of any particle can be determined by the apparatus of Figure 1. As explained in Section 2, this charge is taken by *definition* to be an invariant, independent of both the state of motion, and the coordinate system.

The *electric* field at the point **x** can then be determined by measuring the force $\mathbf{F}(\mathbf{x})$ on a known charge q when it is *held stationary* at the point **x**. Since $\mathbf{v} = \mathbf{0}$, the Lorentz force law tells us that $\mathbf{E}(\mathbf{x}) = \mathbf{F}(\mathbf{x})/q$. Of course, this is precisely the same as our previous definition, given in Section 2.2.

The *magnetic* field at the point **x** is determined from the forces on *moving* charges as they pass through the point **x**. To be explicit, we could proceed as follows. Having found q and $\mathbf{E}(\mathbf{x})$ as above, we allow the test charge to move with velocity **v** along the **3**-axis. Then, as the particle passes through the point **x**, the Lorentz force law tells us that

$$F^1 = q(E^1 + v^2B^3 - v^3B^2) = q(E^1 - v^3B^2)$$
$$F^2 = q(E^2 + v^3B^1 - v^1B^3) = q(E^2 + v^3B^1)$$
$$F^3 = q(E^3 + v^1B^2 - v^2B^1) = qE^3$$

(since $v^1 = v^2 = 0$ as a result of our choice of **v**).

These equations allow us to find two of the components of $\mathbf{B}(\mathbf{x})$:

$$B^1(\mathbf{x}) = \frac{F^2 - qE^2(\mathbf{x})}{qv^3}, \qquad B^2(\mathbf{x}) = -\left(\frac{F^1 - qE^1(\mathbf{x})}{qv^3}\right).$$

The third component of **B** can be found by moving the test charge in another direction, say along the **2**-axis. We would then obtain a result for B^3 as well as a second measurement of B^1.

It is clear that the electric and magnetic fields can readily be unravelled from one another — the electric fields are *defined* with stationary test charges, while the magnetic fields are *defined* from the extra forces that act on moving charges.

(ii)   *As a law*

It cannot be over-emphasized that the Lorentz force law is more than just a *definition* of electric and magnetic fields. It also has the status of a physical law which can be tested by stringent experiments. It is true that **E** and **B** are defined so that the Lorentz force law automatically works for a given test particle both when it is held stationary and when it is moving in a chosen direction. But that does not guarantee that the law should always work. For example, it does not automatically follow that our two measurements B^1 mentioned above should agree. More importantly, however, once **E** and **B** are defined by means of a single test particle which is first held still and then moved in two different directions, the Lorentz force law uses these values of **E** and **B** to predict the forces on *any* particle which is moving with *any* velocity. Clearly, only experiment can decide whether Nature obeys this law — and experiments with currents in wires or beams of charged particles confirm, beyond all question, that the Lorentz force law is correct.

To be precise, we should say that the Lorentz force law correctly describes the forces acting on a particle, and that, in *Newtonian mechanics*, these forces cause the particle to accelerate according to Newton's second law:

$$\mathbf{F} \equiv m\mathbf{a} \equiv m\frac{\mathrm{d}\mathbf{v}}{\mathrm{d}t} = \frac{d}{\mathrm{d}t}(m\mathbf{v}) \equiv \frac{\mathrm{d}\mathbf{p}}{\mathrm{d}t}$$

i.e. the force is equal to the rate of change of momentum of the particle. In Einstein's theory of special relativity, exactly the same statement applies:

$$\mathbf{F} = q(\mathbf{E} + \mathbf{v} \times \mathbf{B}) = \frac{\mathrm{d}\mathbf{p}}{\mathrm{d}t}$$

but, as you will see in Unit 7, the concept of momentum needs to be modified.

Objective 6　　　　SAQ 8　　The velocity $\mathbf{v}$ of an electron can be measured by allowing it to pass through constant electric and magnetic fields which are at right-angles to one another and to $\mathbf{v}$. The strengths of the fields are adjusted until the electron is undeflected.

Determine $|\mathbf{v}|$ if $|\mathbf{E}| = 10^4 \, \mathrm{N\,C^{-1}}$ and $|\mathbf{B}| = 0.01 \, \mathrm{T}$.

3.5　The Lorentz force law and Newtonian mechanics

As we have seen, the magnetic force between two parallel wires depends on the product $|\mathbf{v}_1|\,|\mathbf{v}_2|$ of the speeds of the electrons, rather than the relative velocity $(\mathbf{v}_1 - \mathbf{v}_2)$, and this is in conflict with the Galilean transformation of velocities. This immediately raises doubts about Newton's second law.

At this point, we are presented with two possibilities. One is based on retaining Newton's second law but rejecting the principle of relativity; the other takes the opposite stance and places the prime importance on the principle of relativity. The contrast between these alternatives may be set out as follows:

If we retain Newton's second law

Suppose that two inertial observers, who are related by a uniform linear boost, look at the motion of a single charged particle.

According to the Galilean transformation they must agree about its acceleration:

　　$\mathbf{a}' = \mathbf{a}$　　(Unit 1, Eq. 24).

The invariance of mass means that they also agree about its mass:

　　$m' = m$.

Therefore, by Newton's second law, both observers will find the same force acting on the particle:

　　$\mathbf{F}' = m'\mathbf{a}' = m\mathbf{a} = \mathbf{F}$.

However, the Lorentz force law contains a term with a $|\mathbf{v}_1|\,|\mathbf{v}_2|$ dependence and so the force changes in value from one inertial frame to another.

We therefore conclude that if Newtonian mechanics is to remain unchanged, the Lorentz force law can only hold in one special reference frame. This is tantamount to defining an 'absolute' space and rejecting the principle of relativity.

If we retain the principle of relativity

Like all fundamental laws, the Lorentz force law should take the same form in all inertial frames — otherwise we could use it to distinguish one inertial frame from another.

But two inertial observers whose frames are related by a uniform linear boost will find, in general, that

　　$q(\mathbf{E}' + \mathbf{v}' \times \mathbf{B}') \neq q(\mathbf{E} + \mathbf{v} \times \mathbf{B})$.

Since they agree about the *form* of the Lorentz force, they will disagree about its *value*:

　　$\mathbf{F}' \neq \mathbf{F}$.

Thus if the principle of relativity is retained, Newton's second law has to be sacrificed and his ideas of space and time re-examined.

At the end of the nineteenth century, most physicists believed that Newton's second law was beyond question. However, Einstein challenged this point of view by insisting that the principle of relativity must be retained, *even though* that entailed re-examining Newtonian mechanics and Newtonian concepts of space and time. You will see in Unit 5 how Einstein's approach explained many puzzling experimental results (including the celebrated Michelson–Morley experiment of 1887). Nowadays, his theory is known as 'special relativity' and is universally accepted.

3.6   The field equations for $\mathbf{B}$

Study comment

The comments made in Section 2.4 apply here too. You will not be expected to reproduce or manipulate the field equations written down here.

We have seen that the static $\mathbf{E}$ field is determined by the charge and can be calculated from differential equations in which the charge density acts as a source term. There exist similar equations for $\mathbf{B}$, but as you might guess, they will be different if only because there are no 'charges' for magnetic fields (or, more exactly, the speculated existence of magnetic monopoles has never been confirmed empirically). Hence instead of something analogous to Equation 7,

$$\operatorname{div} \mathbf{E} = \frac{\rho}{\varepsilon_0},$$

we have

$$\operatorname{div} \mathbf{B} = 0. \tag{14}$$

One source of magnetic field is an electric current; if $\mathbf{j}(\mathbf{x})$ is a vector representing the *current density*, i.e. the charge flowing per unit area per unit time at some point $\mathbf{x}$, in the direction of the current flow, then the source equation for $\mathbf{B}$ turns out to be:

$$\operatorname{curl} \mathbf{B} = \mu_0 \mathbf{j}. \tag{15}$$

Thus Equations 14 and 15 are the field equations for $\mathbf{B}$ when currents are the only source of $\mathbf{B}$; the meaning of this qualification will become apparent when another source of $\mathbf{B}$ emerges later.

Once more, we remind you that it is the *form* of these equations you should note; we do not expect you to write them down or do calculations with them in this Course. And do note that they are not yet the final form of these equations; these we present after the next section.

4   Maxwell elucidates light

4.1   Varying magnetic and electric fields

The existence of magnetism has been recognized from very early times, and the electrical properties of amber were explored by Gilbert around 1600. The intimate relationship between them, however, was only revealed by the brilliant and persevering work of Michael Faraday (1791–1867). One of his most far-reaching discoveries was that *when a magnetic field varies with time an electric field is generated.* The easiest way to detect the electric field is via the current which it can cause to flow in a wire loop. The variation in magnetic field can be produced in various ways, for instance by simply moving a bar magnet closer to the loop.

Objective 3   SAQ 9   A bar magnet is moved along the axis of a loop as in Figure 23. The drawing shows the magnetic field lines associated with the magnet. How would you expect the strength of the magnetic field to vary at the position of the loop as the magnet is moved towards the loop?

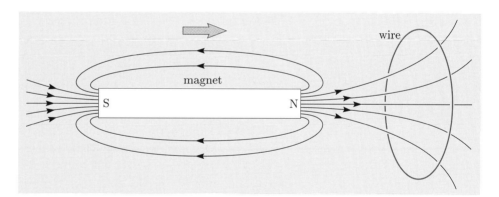

Figure 23   A bar magnet and its associated magnetic field lines. It is being moved nearer to the wire loop along the axis.

The changing magnetic field due to the moving magnet causes an electric field to appear, which is circular around the wire loop. Notice that such an electric field could *never* be created by an arrangement of stationary charges. You saw from Figure 8(a) that a circular electric field would allow us to extract large amounts of energy from the field. If such a field were created by stationary charges, this could not be reconciled with the conservation of energy. But in the case of a moving magnet there is no conflict between circular electric fields and the conservation of energy; for example, the kinetic energy of the electrons in the wire loop comes ultimately from the motion of the magnet. Indeed, to move the magnet against the magnetic forces created by the currents in the loop, energy must be supplied, as in a dynamo.

Objective 3   SAQ 10   A magnetic field can also be produced by a current in a wire. Using Figure 20 as a starting point, sketch the field set up by a single circular loop of wire, in a plane which bisects the loop and is perpendicular to it; show that the pattern of field lines is in some ways similar to that of a bar magnet.

The magnetic field around a loop or coil of wire can easily be varied by changing the current through the wire. Thus if two coils are placed next to each other (Figure 24), a changing current in one produces, via the changing magnetic field and resultant electric field, a changing current in the other. (This is why we restricted ourselves to *steady* currents in Section 3.2. Non-steady currents give rise to electric fields, which would have complicated the issue.)

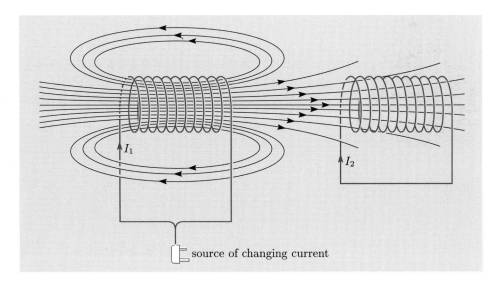

source of changing current

Figure 24   When current I_1 changes in wire 1, it induces a current I_2 in wire 2.

The constants of Nature are such that substantial electric fields, and therefore useful currents, can be produced in these ways. For instance, a modest bar magnet can give rise to a magnetic field of 0.05 T. If such a magnet is snatched away from a loop of wire in (say) 0.1 seconds, the rate of change of the magnetic field is $0.5\,\mathrm{T\,s^{-1}}$, which produces a readily detectable electric field as shown in the video band associated with this Unit.

Faraday's effect   In fact, such effects are at the heart of the electrical industry. Figure 23 represents the fundamental process in a dynamo, and Figure 24 is a rudimentary transformer. For us, however, the importance of Faraday's discovery is that it shows one of the links between electric and magnetic fields: magnetic fields which vary in time, give rise to electric fields which vary in space.

Maxwell's effect   Maxwell guessed that the converse should also be true — an electric field which varies in time, gives rise to a magnetic field which varies in space. This means, for instance, that in the piece of apparatus shown in Figure 25 there could be a magnetic field of the form shown, when an alternating voltage is applied across the plates. This magnetic field looks just like that round a current in a wire, but here there is no current. Even if the space between the plates is a perfect vacuum, the field still exists.

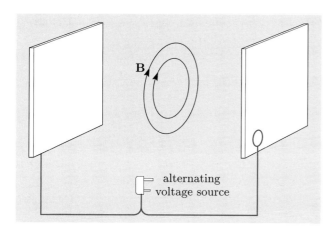
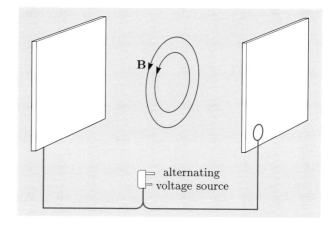

Figure 25   Two snapshots of the magnetic field produced by an alternating electric field between two plates

Figure 26   James Clerk Maxwell 1831–1879.

Nowadays, it is possible, though tricky, to perform an experiment of this type and confirm Maxwell's intuition was right. But in Faraday's day, the experiment was not feasible. This was because the constants of Nature are such that the magnetic fields are very small. For instance, if the voltage on the plates varies between $+10\,000$ V and $-10\,000$ V at $10\,000$ hertz and the plates are 0.05 m apart, the magnetic field is only about 10^{-9} tesla. This is less than a millionth of the magnetic field produced by a small bar magnet, and would not have been detectable in Faraday's time. Nevertheless, Maxwell (Figure 26) felt convinced that, for the sake of symmetry, this small effect must exist. In fact it was this insight that allowed him to complete his theory of electromagnetism, with equations which were soon to show their elegance and power.

4.2   Maxwell's equations

Study comment

The comments made in Section 2.4 apply here too. You will not be expected to reproduce or manipulate the field equations written down here.

We have now seen that there are new sources of both **B** and **E** fields: a changing **B** field (i.e. one for which the time dependence $d\mathbf{B}/dt \neq 0$) is a source of **E**, and a changing **E** field is a source of **B**. This new physics is represented mathematically by adding new source terms to two of the four equations, i.e. Equation 6 and Equation 15 which become:

$$\operatorname{curl}\mathbf{E} = -\frac{\partial \mathbf{B}}{\partial t} \tag{16}$$

$$\operatorname{curl}\mathbf{B} = \mu_0 \mathbf{j} + \varepsilon_0 \mu_0 \frac{\partial \mathbf{E}}{\partial t}. \tag{17}$$

For completeness, we restate the other two equations, which are unchanged:

$$\operatorname{div}\mathbf{E} = \frac{\rho}{\varepsilon_0} \tag{18}$$

and

$$\operatorname{div}\mathbf{B} = 0 \tag{19}$$

Maxwell's equations

(where Equation 18 is Equation 7 and Equation 19 is Equation 14).

These four equations are known as Maxwell's equations for free space, and represent a *complete* theory of the classical electromagnetic field in free space (free space, that is, apart from the sources). Together with the Lorentz force law, which says what we mean by **E** and **B** in terms of forces, they constitute classical electromagnetism; needless to say, drawing out the consequences of these equations in practical circumstances is a huge and still developing subject.

4.3 Electromagnetic waves

What really crowned Maxwell's achievement was his demonstration that the four equations implied the existence of electromagnetic *waves*. This comes about because a changing **E** field produces a **B** field (see Equation 17). Moreover, that **B** field will also be changing, so it is, in turn, the source of a changing **E** field, which then creates a changing **B** .... In other words, the possibility arises of a mutually self-sustaining set of changing **E** and **B** fields. Moreover, Maxwell found that these coupled electrical and magnetic disturbances would travel through space with a speed which, in customary modern notation is

$$c = \frac{1}{\sqrt{\varepsilon_0 \mu_0}}. \tag{20}$$

Since, in effect, ε_0 and μ_0 were fixed by static experiments involving electric and magnetic fields, one can only imagine Maxwell's feelings when he evaluated c from this expression and lo! it was just the speed of light measured by Foucault, Fizeau and others. Notice that in free space where **j** and ρ are both zero, μ_0 and ε_0 only occur in Maxwell's equations as the product $\mu_0 \varepsilon_0$ and this is reflected in Equation 20.

Of course, the process whereby light was generated was, in Maxwell's day, obscure. From the measured wavelengths, the deduced frequency of visible light, f $(= c/\lambda)$ seemed unimaginably high, some 10^{15} Hz. If this led to any doubts that Maxwell had found the nature of light, these were dispelled by the experiments of Heinrich Hertz in 1887. Hertz generated waves in the laboratory from oscillating electric currents in circuits, and then showed that these waves had *all* the properties of light, but with frequencies of some 10^8 Hz. From this followed the development of radio by Marconi and others; it also left any lingering adherence to 'instantaneous action at a distance' stone dead.

We now understand that all forms of electromagnetic radiation, from long wavelength radio waves to gamma rays of wavelengths comparable to the dimension of atomic nuclei, are wave solutions of Maxwell's equations, and all travel at $c = 299\,792\,458\,\mathrm{m\,s^{-1}}$. (The lack of any quoted uncertainty in c is not a slip; that value is now the exact accepted value incorporated into the SI system of standards of length and time measurements.)

Maxwell's equations are now a deeply embedded ingredient in all of modern physics. Unfortunately, it is just outside the scope of this course* to show mathematically how waves arise as solutions to these equations. But we can say a bit more, qualitatively, concerning why it is that their speed is related to the fundamental constants ε_0 and μ_0, and, in the next section, how they are generated. Figure 27(a) shows a disturbance in the electric

*but not of SMT356, *Electromagnetism*

field, travelling with uniform velocity along the **2**-axis. For simplicity, we assume that the electric field is directed only along the **1**-axis.

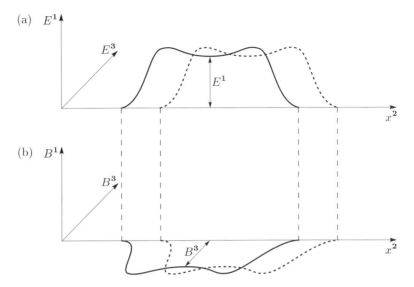

Figure 27 Disturbances in (a) the electric and (b) the magnetic fields propagating along the x^2-direction

As this pulse moves, the electric field at a given point changes with time. According to the ideas of Maxwell (particularly that encapsulated in the arrangement of Figure 25), this means that the moving electric pulse is accompanied by a magnetic pulse at right angles to both **E** and the direction of motion of the disturbance; this is shown in Figure 27(b). The strength of this magnetic field depends on two factors:

(a) the rate of change of the electric field, which, in turn, depends on the speed with which the electric pulse is moving, and

(b) the constants μ_0 and ε_0, which tells us how much magnetic field is produced by a given electric field (cf. SAQ 6).

As the magnetic pulse moves, the magnetic field at a given point changes in time and so, according to Faraday's effect, it must be accompanied by an electric field which varies in space. Hence, the moving magnetic pulse is automatically associated with an electric pulse whose size depends on the size and speed of the magnetic pulse.

It is here that we meet a self-consistency requirement. **Since the electric field at a given place and time must have a well-defined value, the electric field required by a combination of Maxwell's and Faraday's effects must be identical to the electric field we started with.**

This means that, for given values of the constants ε_0 and μ_0, there is only *one* speed at which the electric pulse can propagate through space. Imagine, for example, an electric pulse which is travelling too slowly. This pulse would give rise (via Maxwell's effect) to a weak and sluggish magnetic pulse. But, according to Faraday's effect, such a magnetic pulse can only be accompanied by a *weak* electric pulse — weaker than the pulse we originally envisaged. Such a state of affairs is clearly impossible. It is only if the electric pulse moves at exactly the right speed that the spatial and temporal changes in the electric field can be consistent with *both* Faraday's and Maxwell's effects. We conclude that, for given constants ε_0 and μ_0, there is only one speed, c, at which electric pulses can move through empty space.

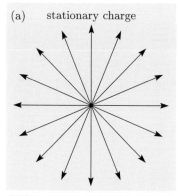

(a)   stationary charge

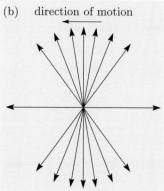

(b)   direction of motion

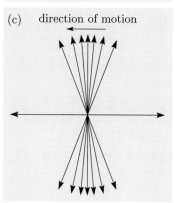

(c)   direction of motion

Figure 28 The electric field lines of (a) a stationary charge, (b) a charge moving slowly, and (c) a charge moving with a greater speed than that in (b).

How are we to create pulses of electric and magnetic fields which move at speed c? It turns out that if we *accelerate* a charge, we disturb the fields around it, and this gives rise to disturbances which move with the necessary speed. In the next section, we shall give a brief description of this process of wave generation.

4.4   The generation of electromagnetic waves

Figure 28(a) shows the familiar pattern of electric field lines characteristic of an electric charge at rest. What pattern exists around a charge moving with a uniform velocity? Maxwell's equations allow the pattern to be calculated. One such pattern is shown in Figure 28(b). As you can see, the field is no longer spherically symmetric; the field lines become concentrated at progressively greater angles to the direction of motion. This effect becomes more and more pronounced the greater the speed of the particle carrying the charge, as is illustrated in Figure 28(c). The field still exhibits a symmetry, but now it is only an axial symmetry about the direction of motion.

So much for charges that are stationary or in a state of uniform motion. What of a charge that undergoes an acceleration or deceleration? Clearly we have to switch from one pattern of field lines to another. For example, a particle originally moving uniformly and then suddenly brought to rest must change the pattern of field lines appropriate to its original motion (a pattern similar to Figures 28b or 28c) to the one characteristic of the stationary charge (Figure 28a). How is this to be done? Would the change take place instantaneously? If it did, then a distant test charge would immediately find itself in the new field and the information that the source charge had now come to rest would have been communicated to it instantaneously. This would be an example of instantaneous action-at-a-distance.

Not surprisingly, an instantaneous changeover from one pattern of field lines to the other does *not* happen. In reality, the change is made progressively. It is only after a certain period of time that the distant test charge experiences the new field, the time delay depending on the distance between it and the source charge. The situation is illustrated in Figure 29. From this diagram, we can see that the field lines undergo a sharp lateral movement in order to take up their new configuration. It is this lateral 'kink' that moves progressively outwards.

An electric pulse, however, is not the only thing travelling outwards from the decelerating charge. The electric pulse represents a *change* in the electric field, and we have learnt that a change in an electric field is invariably accompanied by a magnetic field. The pulse therefore has not only an electric component but also an associated magnetic component — it is an *electromagnetic disturbance*.

Now we can answer the following question: are fields 'real'? In other words, when we introduce the idea of a field **E** as an alternative to the action-at-a-distance which Coulomb's law suggests, is **E** *simply* a mathematical concept that simplifies some calculations? Alternatively, is there something *real* filling space near a charge? Field lines have been a handy reminder of the strength and direction of the forces a test charge would experience if it were placed in the vicinity of the source charge. It is with the study of the lateral pulses illustrated in Figure 29 that the physical reality of fields becomes clearer. A test charge placed in the path of such a pulse really would receive a sudden sideways kick. The kick would give it energy and momentum, both coming originally from the motion of the source charge but now being carried by the pulse itself. Anything carrying energy and momentum must be physically real. We therefore conclude that fields are more than mathematical concepts, they are part of the fabric of the world.

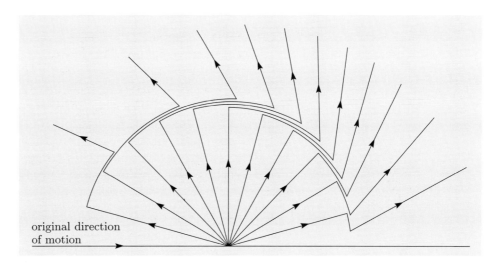

original direction
of motion

Figure 29   The electric field lines of a charge that has been moving at a constant velocity and is abruptly brought to rest. Information that the charge has stopped has not yet reached beyond the hemisphere. Outside this zone the lines are those characteristic of the moving charge and point to where that charge would have reached had it continued moving.

This is illustrated very clearly in the act of taking a photograph. The chemical changes in the film need energy to make them occur, and it is the light that brings this energy. The same is true of photographs taken with X-rays. The Sun's infrared radiation also gives us energy in the form of heat, and its ultraviolet radiation causes chemical changes in our skin and maybe gives us a tan. Thanks to Maxwell, we now understand that all these effects occur because the electric and magnetic fields are real — real enough to carry energy and momentum across vast regions of space.

Another remarkable feature of electromagnetic waves is that they can travel through a vacuum. The radiations from the Sun and stars reach us only after having passed through almost empty space. In this respect, they differ from other wave forms which require a material medium. For example, sound waves need air or perhaps the ceiling between you and the flat upstairs; water waves, of course, need water.

The speed of sound waves depends on the properties of the medium carrying them — its density and compressibility. Similarly, the speed of (shallow) water waves depends on the density and surface tension of the water.

The speed of light, however, depends only on the two constants ε_0 and μ_0, and these describe the nature of *empty space*. Empty space does indeed have properties, and we shall see in Block 3 how these properties can even be changed by the presence of matter. The constants ε_0 and μ_0 describe the way that empty space responds to, and links together, electric and magnetic fields. The speed of light, in combining these two constants, reflects in a very deep way the nature of space in our Universe.

4.5   Reference frames — a problem

We saw in Section 3.5 how the form of the Lorentz force, implying as it does a force between moving charges depending on $|\mathbf{v}_1||\mathbf{v}_2|$, presents a problem. A force of that form is not Galilean invariant. Stated another way, in what frame is the $\mathbf{v}$ in Equation 11 to be evaluated? The same kind of problem now arises over Maxwell's equations. The unification of electric and magnetic phenomena within a common theoretical framework and the prediction of electromagnetic waves was an enormous achievement. But to which observer did the predicted value of the speed of light refer? We know that two observers in relative motion, when viewing some third object, will have different descriptions of its velocity. In Newtonian space and time, the two values will be related by a straightforward addition, called the Galilean transformation. But now suppose the object they are viewing is a light pulse. For whom will the Maxwellian value of $(\varepsilon_0\mu_0)^{-1/2}$ apply?

As before, we are clearly in a dilemma. The numbers ε_0 and μ_0 are universal constants — an integral part of the laws of electric and magnetic forces. If we believe in the principle of relativity, then we expect all observers in uniform relative motion to have the same laws of physics. In particular, their experiments on electric and magnetic forces must yield the same values of ε_0 and μ_0. But aren't Maxwell's equations also to be regarded as 'laws of physics' and so shouldn't both observers have a right to expect the speed of light to have the same value $(\varepsilon_0\mu_0)^{-1/2}$? Such a conclusion would violate the Galilean transformation.

Earlier, when considering the Lorentz force law, we decided that we could go one of two ways: either accept the principle of relativity and the universal applicability of the Lorentz force law to all inertial observers, and reject Newtonian mechanics and the Galilean transformation; or, alternatively, accept the transformation as it is and abandon the principle of relativity, at least as far as the universal applicability of the Lorentz force law is concerned. Now we are faced with a similar situation over the applicability of Maxwell's equations. Their universal applicability to all inertial observers as required by the principle of relativity would necessitate a revision of the Galilean transformation in order to preserve the universality of the speed of light.

Essentially, the two dilemmas are different ways of expressing the same problem. The Lorentz force law is closely related to Maxwell's equations, so if there is something 'odd' about it, this will reflect itself in the working of Maxwell's equations. In the next Unit, you will see how either approach leads to the theory of special relativity.

Summary

1 Electric charge is a fundamental property of matter. We assume charge is invariant under coordinate transformations, and total charge is conserved.

2 The magnitude of an electric charge can be defined, in a practical way, as being proportional to the force on it in an electric field that has been generated in a specified way.

3 The force $\mathbf{F}_{12}$ on stationary charge q_1 due to the presence of stationary charge q_2 is given by Coulomb's law

$$\mathbf{F}_{12} = \frac{1}{4\pi\varepsilon_0}\frac{q_1 q_2}{x^2}\mathbf{e}_{12},$$

where x is the distance between the charges, ε_0 is a universal constant and $\mathbf{e}_{12}$ is a unit vector in the direction from $\mathbf{x}_2$ to $\mathbf{x}_1$. Unlike charges experience an attractive force along the line between them; like charges repel each other.

4 There is a striking parallel between Coulomb's law and the inverse square law of gravitational forces. (Note that charge, like mass, is an unexplained fundamental property.)

5 While Coulomb's law is true for static charges, the picture it suggests, namely, instantaneous action-at-a-distance, does not generalize to moving charges. The alternative formulation of electrostatics involving an electric field, $\mathbf{E}$, permits a purely *local* description of electrostatic forces.

6 Field lines give the direction of the force, and their density can be related to the strength of the field. There are severe restrictions on the shapes of possible fields. These arise from the need to avoid the possibility of extracting unlimited energy from the field, and by the requirement for continuity.

7 When two charges are in motion, an additional force between them appears over and above the electrostatic force. This is again well described by a field: a magnetic field generated by the motion of one charge and acting on the other moving charge. An example of this is the force between two parallel, current-carrying wires, which depends on the product of the currents.

8 Both charges must be in motion for the force to exist. The strength of the force as measured by an observer depends on the velocities of the charges relative to the observer, rather than their velocities relative to each other. This fact highlights a conflict with Newtonian mechanics, and in Unit 5 it will allow us to see directly into the workings of relativistic mechanics.

9 The combined effect of electric and magnetic fields is given by the Lorentz force, $\mathbf{F} = q(\mathbf{E} + \mathbf{v} \times \mathbf{B})$.

10 Electric and magnetic fields are linked by their time dependence — a varying magnetic field produces an electric field. Maxwell predicted, correctly, that the inverse is also true: a changing electric field generates a magnetic field.

11 Maxwell found a set of four differential equations which completely describe how the $\mathbf{E}$ and $\mathbf{B}$ fields vary in space and time. Each equation has a spatial differential of $\mathbf{E}$ or $\mathbf{B}$ on the left-hand side and a source term on the right-hand side. The sources of the fields are charges and currents, and also time-dependent fields.

12 By the use of field lines, we were able to show that an electric charge that undergoes acceleration (or deceleration) produces an electric disturbance which spreads out in space. This change in the electric field must be accompanied by a magnetic field, so the resultant disturbance is called an electromagnetic wave.

Band 2 of AC2 comments on this Unit.

13 Maxwell showed from his equations that electromagnetic waves have a speed, c, given by $(\varepsilon_0 \mu_0)^{-1/2}$, where ε_0 and μ_0 are constants which, in effect, can be determined by experiment. Thus the speed of the wave is independent of its frequency and wavelength. This speed has the same value as that of light. Light is therefore taken to be one example of an electromagnetic wave.

14 The electromagnetic wave carries energy and momentum (and may also carry net angular momentum). The fact that these properties can be carried through empty space by the wave establishes the physical reality of fields **E** and **B**, even though they were introduced initially as a convenient description.

15 As was the case with the Lorentz force law, Maxwell's equations appear to conflict with Newtonian physics. This can be seen by asking to which observer the calculated speed of light applies.

Acknowledgements

Grateful acknowledgement is made to the following sources for material used in this Unit:

Figure 22 from the Science Museum, London; *Figure 26* from the Mansell Collection.

Self-assessment questions — answers and comments

SAQ 1 Coulomb's law gives the magnitude of the force as

$$F = 9 \times 10^9 \times \frac{q_1 q_2}{r^2}$$

$$= 9 \times 10^9 \times \frac{(1.6 \times 10^{-19})^2}{(10^{-15})^2} = 230 \, \text{N}.$$

This enormous repulsive force is overcome by the attractive nuclear force.

The electric field is defined as the force per unit charge on the test particle. Regarding one of the protons as a test particle of charge $q = 1.6 \times 10^{-19} \, \text{C}$, the field is

$$\frac{230}{1.6 \times 10^{-19}} = 1.44 \times 10^{21} \, \text{SI units.}$$

SAQ 2 Figure 30 shows a point A that is equidistant from the two charges Q_1 and Q_2 of Figure 5. Consider a set of four equal positive test charges labelled q_1 to q_4 placed at A. The forces on these test charges are as shown in Figure 31, for the following reasons:

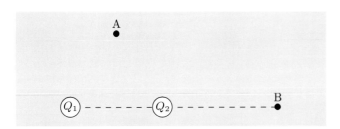

Figure 30 Charges $Q_1 = -3$ and $Q_2 = +3$, equidistant from a point A and aligned with a point B.

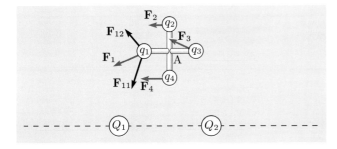

Figure 31 The forces acting on four equal, positive test charges q_1–q_4 situated close to the point A, due to Q_1 and Q_2.

The charge q_1 is attracted by Q_1 with a force $\mathbf{F}_{11}$ and it is repelled by Q_2, with a force $\mathbf{F}_{12}$. (Notice that $|\mathbf{F}_{12}| < |\mathbf{F}_{11}|$ because q_1 is further from Q_2 than from Q_1.) The total force acting on q_1 is found using the superposition of forces: $\mathbf{F}_1 = \mathbf{F}_{11} + \mathbf{F}_{12}$.

Repeating this procedure for q_2, q_3 and q_4, we arrive at the net forces shown in blue in Figure 31. The forces on q_2 and q_4 together give rise to a **resultant** tendency to twist the set of four test charges clockwise. The forces on q_1 and q_3 are equal in magnitude but have different directions. They also

give rise to a tendency to twist the set of test charges, but this time in the anticlockwise sense. Thus these two tendencies are in opposite senses and, in fact, will cancel out. This then is a *possible* configuration.

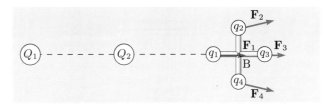

Figure 32 The forces acting on four test charges q_1–q_4 situated close to point B, due to Q_1 and Q_2.

A similar situation applies at B (see Figures 30 and 32). The forces on q_2 and q_4 are equal, but tend to turn the system anticlockwise and clockwise respectively and therefore cancel out. The forces on q_1 and q_3 pass through B, and therefore can have no tendency to turn the set of test charges. A similar argument would apply to *any* point along the broken line in Figure 30.

Points A and B are, of course, particularly simple points to consider and were chosen for that very reason. But in general, for any point in the field, a tendency on the part of one pair of test charges to twist the cross one way will be counterbalanced by opposing forces acting on the other two charges and tending to twist the cross the other way.

There is, of course, a translational force on the set of four charges in all cases, because together they make up a test charge, of finite but arbitrarily small size, which will accelerate in the direction of the field lines.

SAQ 3 Newton's law of gravitation is very similar in form to Coulomb's law, except that masses take the place of charges. We could define a *gravitational* field in terms of the *gravitational* forces on a test particle and then proceed exactly as in Section 2.3 of this Unit, deriving a result analogous to Equation 5.

Just as for an electric field, the average radial gravitational field over the imaginary sphere S in Figure 33 depends only on the amount of matter contained within S (and *not* on the way this matter is distributed). It would therefore make no difference to the average radial gravitational field on S if the Earth were shrunk to a point at the centre of S.

Moreover, assuming the Earth is spherical, its gravitational field is always radial (perpendicular to S) and has the same magnitude at each point on S. So taking the average radial component makes no difference — the average radial field is the same as the field at any point on S. We conclude that, at any point outside its surface, the Earth behaves as if all its mass were concentrated at its centre. This result is of great historical importance. It also illustrates how powerful the idea of a field is. It took Newton *many years* to arrive at the same conclusion because it presented him with a more difficult mathematical task, for which he had to invent integral calculus.

Figure 33   A sphere whose centre is at the centre of the Earth.

SAQ 4   According to Equation 5, the area of a sphere, multiplied by the average radial field at its surface, is proportional to the total amount of charge contained within the sphere. Since the wire carries a uniform distribution of charge, it follows that

$$\frac{(4\pi r_1^2)(\text{average radial field on sphere 1})}{(4\pi r_2^2)(\text{average radial field on sphere 2})} = \frac{2r_1\lambda/\varepsilon_0}{2r_2\lambda/\varepsilon_0}$$

where λ is the charge per unit length of wire.

Hence $\dfrac{(\text{average radial field on sphere 1})}{(\text{average radial field on sphere 2})} = \dfrac{1/r_1}{1/r_2}$.

So the average radial field on a given sphere is inversely proportional to its radius.

SAQ 5   Suppose that the first wire contains n_1 electrons per unit length, each moving with velocity $\mathbf{v}_1$. These electrons give rise to a total current of magnitude $I_1 = n_1 e|\mathbf{v}_1|$ (where e is the magnitude of the electron charge). The second wire contains n_2 electrons per unit length, each moving with velocity $\mathbf{v}_2$. These electrons give rise to a total current of magnitude $I_2 = n_2 e|\mathbf{v}_2|$. The current in the first wire produces a magnetic field in the vicinity of the second, whose size is

$$|\mathbf{B}| = \frac{\mu_0}{4\pi}\frac{2I_1}{d}.$$

This magnetic field acts on each of the electrons in the second wire with a force

$$\mathbf{F} = q\mathbf{v}_2 \times \mathbf{B},$$

where $q = -e$ so the magnitude is $|\mathbf{F}| = e|\mathbf{v}_2|\,|\mathbf{B}|$ since $\mathbf{v}_2$ and $\mathbf{B}$ are at right-angles to each other.

Thus the *total* force acting on the second wire due to the current in the first wire is in the direction $\mathbf{v}_2 \times \mathbf{B}$ (i.e. it is perpendicular to the two wires as

expected), and it has magnitude

$$|\mathbf{F}| = n_2 e|\mathbf{v}_2|\,|\mathbf{B}| = \frac{\mu_0}{4\pi}\frac{2I_1 I_2}{d}$$

in agreement with Equation 10.

SAQ 6   **(a)**   When a steady electric current flows in a wire, experiments show that no electric field is produced outside the wire. The electric field $\mathbf{E}_e$ due to the moving negative electrons must therefore exactly cancel the electric field due to the stationary positive ions.

Since the field due to the ions is, presumably, independent of the velocity of the electrons, it follows that the combined electric field due to all the electrons in the moving stream cannot depend on their rate of flow.

(b)   The strength of the current due to the moving electrons, with magnitude of charge e, is $ne|\mathbf{v}|$, and this gives a magnetic field

$$|\mathbf{B}| = \frac{\mu_0}{4\pi}\frac{2I}{r} = \frac{2|\mathbf{v}|ne}{r}\left(\frac{\mu_0}{4\pi}\right).$$

Writing ne/r in terms of $\mathbf{E}_e$, we find

$$|\mathbf{B}| = 2|\mathbf{v}|\frac{4\pi\varepsilon_0|\mathbf{E}_e|}{2}\left(\frac{\mu_0}{4\pi}\right)$$
$$= \varepsilon_0\mu_0|\mathbf{v}|\,|\mathbf{E}_e|.$$

The electric field is radial, and $\mathbf{v}$ is along the wire. The magnetic field is tangential to a circle around the wire. This relationship between the directions is correctly reflected by writing

$$\mathbf{B} = \varepsilon_0\mu_0(\mathbf{v} \times \mathbf{E}_e),$$

with the direction of $\mathbf{B}$ as in Figure 20.

SAQ 7 $\mathbf{B} = (2, 3, 5)$ so $B^1 = 2, B^2 = 3, B^3 = 5$;
$\mathbf{v} = (1, 2, 3)$ so $v^1 = 1, v^2 = 2, v^3 = 3$.

$$\begin{aligned}\mathbf{v} \times \mathbf{B} &= (v^2 B^3 - v^3 B^2, v^3 B^1 - v^1 B^3, v^1 B^2 - v^2 B^1)\\ &= [(2\times5 - 3\times3), (3\times2 - 1\times5), (1\times3 - 2\times2)]\\ &= (1, 1, -1).\end{aligned}$$

So the force on a unit charge $= 1 \times (1, 1, -1)$
$$= (1, 1, -1) \text{ newtons.}$$

The magnitude of this force is $\sqrt{1^2 + 1^2 + 1^2} = \sqrt{3}\,\mathrm{N}$. The maximum magnitude of the force is $|\mathbf{v}|\,|\mathbf{B}|$ and is found when $\mathbf{v}$ and $\mathbf{B}$ are at right-angles (when $|\mathbf{v}||\mathbf{B}|\sin\theta = |\mathbf{v}||\mathbf{B}|$).

$$|\mathbf{v}| = \sqrt{1^2 + 2^2 + 3^2} = \sqrt{14};$$
$$|\mathbf{B}| = \sqrt{2^2 + 3^2 + 5^2} = \sqrt{38};$$
$$\therefore\ |\mathbf{v}|\,|\mathbf{B}| = \sqrt{532} \approx 23.1.$$

SAQ 8   The combined force is

$$\mathbf{F} = q(\mathbf{E} + \mathbf{v} \times \mathbf{B}).$$

If $\mathbf{E}$, $\mathbf{B}$ and $\mathbf{v}$ are at right-angles to each other ('mutually orthogonal') and the total force is zero,

we have

$$\mathbf{E} = -\mathbf{v} \times \mathbf{B}$$
$$|\mathbf{E}| = |\mathbf{v}|\,|\mathbf{B}|$$
$$|\mathbf{v}| = |\mathbf{E}|/|\mathbf{B}| = 10^4/0.01 = 10^6.$$

Thus the speed is $10^6\,\mathrm{m\,s^{-1}}$.

SAQ 9 When the bar magnet is far away from the loop of wire, the magnetic field is weak and directed at a small angle to the axis of the magnet. As the bar magnet approaches, the field gets stronger and also the angle increases. After the bar magnet has entered the loop, the direction of the field reverses.

SAQ 10 Consider the circular lines of force as the wire is gradually bent into the loop. They crowd together at the centre of the loop, which means the field is stronger there than outside the loop. The lines in a plane through the loop look as in Figure 34.

If the central section of this pattern is blocked out, the rest is similar in shape to that shown in Figure 23.

Figure 34 The field produced by a current in a circular loop of wire, in a plane that bisects the loop and is perpendicular to it.

Unit 5 The need for special relativity

Prepared by the Course Team

Contents

Aims

In this Unit, we intend to:

1 Show, by considering two or three 'thought-experiments', that if the new Maxwell–Faraday theory of electromagnetism is accepted as a good model of certain natural phenomena, then the old Newtonian attitudes to space and time need modifying.

2 Introduce Einstein's (1905) modification of the Newtonian world-view, which is suitable when gravity can be neglected.

3 Describe the definition of time appropriate to the Einstein world-view, for any given inertial frame.

4 Reconsider spacetime diagrams and, using them, begin a fairly systematic discussion of Einstein's 1905 theory of special relativity which will be continued throughout the remainder of the Block.

Objectives

When you have finished studying this Unit, you should be able to:

1 Describe the thought-experiment of Section 2, listing the assumptions and conclusions.

2 Define the 'luminiferous ether', and explain why its existence seemed possible or even necessary to many physicists in the latter part of the nineteenth century.

3 Explain how certain features of the new Maxwell electromagnetism conflicted with the old Newtonian world-view.

4 State Einstein's two postulates of relativity and explain at some length what they are intended to say.

5 Describe a thought-experiment with and without using spacetime diagrams, to show that two events that are simultaneous relative to one inertial observer are not, in general, simultaneous for another inertial observer moving relative to the first.

6 Give two equivalent operational definitions of time for any chosen inertial frame.

7 Using these definitions, explain what we mean by 'observer' and by 'spacetime diagram'.

8 Describe the Michelson–Morley experiment in the spirit of Appendix A and explain its relevance to late nineteenth century physics.

9 Outline briefly, without going into details, the experiments listed in Appendix B which showed that the speed of light is independent of frequency, time or place of measurement, the nature or motion of the source, and the inertial frame from which the measurement is made. Explain why such results rule out the ballistic theory of light.

1   Introduction to Unit 5

Band 3 of AC2 introduces this Unit.

As you know from Unit 4, the new subject of electromagnetism (or electrodynamics) considered certain natural phenomena not easily explained by Newtonian mechanics. For instance, how does one easily explain the currents induced in a coil by moving magnets without the new concept of electromagnetic fields? Then there is what might be called 'the problem of light'. From the time of Newton himself, many varied speculations and theories had been devised to explain equally varied properties of light; examples are the speed of propagation of light, its wave-like properties manifested in certain refraction and interference experiments, and its ability to propagate through virtually *empty* space from distant stars quite in contrast to other known waves which require a *medium* in which to move as do water waves and sound waves.

Nowadays, we appreciate that visible light is just electromagnetic radiation, but the various seemingly contradictory properties of light provided perhaps the major motivation to physicists of the latter part of the nineteenth century towards the modern theory of electrodynamics — which encompasses particle dynamics on the one hand and the dynamics of electric and magnetic fields on the other. The synthesis came in 1905 with Albert Einstein's great paper 'Zur Electrodynamik bewegter Körper' (i.e. 'On the electrodynamics of moving bodies'; *Annalen der Physik*, **17**, 891), wherein he showed that Maxwell's equations obeyed the principle of relativity provided that (i) the Galilean transformation of coordinates (Unit 1, Section 5.3) is replaced by a new transformation, the Lorentz transformation (Unit 6, Section 2.3), and (ii) Newton's mechanics are suitably altered.

We shall be able to consider some of the content of Einstein's 1905 paper in Units 5–7. We have chosen *not* to approach the modern theory of special relativity by the usual strictly historical path, which would consider the development of theories about the nature of light from the eighteenth century to the time of Maxwell, Faraday and Einstein. Many standard treatments of special relativity do impart a carefully ordered historical flavour to their exposition.

In accordance with our approach as outlined in the Introduction to Block 2, we begin, in Section 2, with an argument based partly on experimental evidence and partly on some not unreasonable hypotheses. This argument throws serious doubt on the Newtonian Assumption A10 of Block 1 that the length or distance between two points in space is the same for all observers, and follows the spirit (we hope) but not specific content, of Einstein's 1905 paper. In Section 2.3 of Unit 6, when we have the explicit formulation of Einstein's theory to hand, we show that one of the basic assumptions (the assumption of charge invariance) is consistently treated by Einstein's theory. The argument is somewhat long, but this is not surprising when one considers that it sets out to undermine 200 years and more of 'obvious' assumptions.

The main purpose of the argument of Section 2 is to cast doubt on Newton's world-view, and we suggest that the length and somewhat intuitive nature of the argument may seem somewhat less formidable if it is reread later, perhaps even after Unit 6 has been completed.

2 A thought-problem: implications for the measurement of length

Let us accept certain experimental results from electrodynamics. In particular, let us accept as experimentally verified the interaction between fields **E** and **B** on the one hand and a charged particle on the other, in the form of the following equation:

$$\mathbf{F} = q(\mathbf{E} + \mathbf{v} \times \mathbf{B}). \tag{1}$$

This equation was considered in Section 3.4 of Unit 4. In Equation 1, **F** is the vector force acting on a particle with velocity **v** and charge q (which may be positive, as for a proton, or negative, as for an electron). The force arises from the combined action of the electric field **E** and the magnetic field **B**. In general, we must suppose that fields **E** and **B** depend on the position (x^1, x^2, x^3) and on time t, but for the present argument we shall consider only time-independent fields such as the magnetic field produced by a *steady* (time-independent) current through a wire. That the force on a charge q can be described by such a simple form as Equation 1 is a law of Nature and therefore is experimentally verifiable. Simple form notwithstanding, to assume Equation 1 is valid is a large assumption. As we stressed in Unit 4, it conflicts with some fundamental ideas of Newtonian mechanics. In the Newtonian world-view, force equals mass times acceleration, and neither mass nor acceleration changes as one goes from one inertial frame to another moving uniformly relative to the first. But in Equation 1, we have a force that depends on *velocity*, and velocity *does* change under such a transformation. Clearly there is trouble brewing!

It is easy to establish a steady current through a wire (of copper, say) by means of a battery. If the wire is long and straight and if the current settles down to some constant value, we have the situation depicted in Figure 1(a) somewhere along the wire's length.

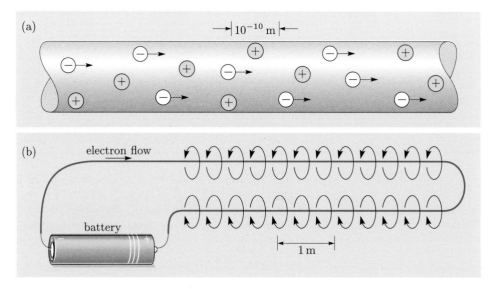

Figure 1 (a) The distribution of fixed positive ions and moving electrons in a section of current-carrying wire.

(b) The magnetic field produced by a current-carrying wire. By convention, the direction of the current is opposite to that of the electron flow shown by arrows.

In Figure 1, we have tried to convey the following experimentally verifiable facts:

1 In the process of conduction of charge through the (copper) wire, only the electrons in the wire are actually mobile. Most of the mass of the (copper) wire is contained in the (copper) ions and these are fixed in the lattice of the wire. In general, the transport of charge through a wire is effected by the motion of electrons. In Figure 1(a), the electrons (whose charge is always negative) are represented by moving minus signs, while the positive copper ions are indicated by fixed plus signs.

2 The diameter of a copper ion is 10^{-10} m, give or take a factor of two.

3 A steady magnetic field is established outside the wire, indicated by circles in Figure 1(b).

If we consider such a wire on a microscopic scale as in Figure 1(a), it is clear that it is very inhomogeneous or grainy. However, let us consider an actual wire such as one might connect to a battery. If the wire has a typical cross-sectional area of $1\,\text{mm}^2$ ($= 10^{-6}\,\text{m}^2$) and carries a current of 10 amperes, it can be shown that about 6×10^{19} electrons per second are flowing through the wire. Furthermore, if all these electrons are flowing uniformly along in a straight line, their mean speed is only something like $1\,\text{mm s}^{-1}$ ($= 10^{-3}\,\text{m s}^{-1}$). Thus in a rather heavy wire carrying a rather heavy current, the current-carrying electrons are drifting along at a very low rate. Surely Newtonian concepts of space and time ought to apply here? We shall see that they do not!

We now argue that in the laboratory, any relevant dimensions are so *very much* bigger than any dimension relating to a single electron that we may think of the conduction process as being uniform and homogencous. Let us therefore model the situation, as in Figure 2, by a uniform line of electrons moving off to the right with a 'drift velocity' **u**.

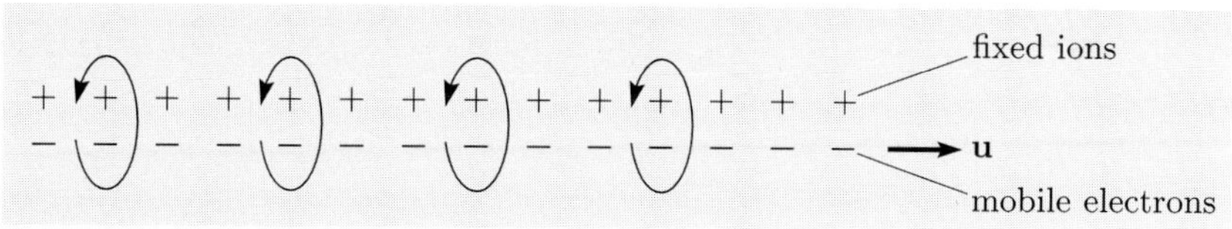

Figure 2   Laboratory frame

How can we determine the electric and magnetic fields produced by the wire? The general method was described in Section 3.4 of Unit 4, but to remind yourself of it look again at the Lorentz force law (Equation 1). Applying this equation, still *in a reference frame fixed to the laboratory* (called the 'laboratory frame'), we see that to measure the electric field **E** outside the wire one merely has to measure the force on a test charge q that is motionless, i.e. with $\mathbf{v} = 0$. When the experiment is done, it turns out that the electric field is zero. In other words, *the force on a test charge lying stationary on the outside of a wire carrying a* steady *current is zero*. Having established **E** to be zero in the laboratory frame, now consider the magnetic field. One can verify experimentally that the field **B** outside a long wire carrying a *steady* current, is directed in concentric circles as in Figures 1(b) and 2. The magnitude, $|\mathbf{B}|$, decreases with distance away from the wire, but we don't need to pursue that here.

Let us now consider the situation in detail from two different frames of reference. For definiteness, let us choose a negatively charged electron ($q < 0$) for our test particle. If we suppose that the electron moves uniformly to the right (in the laboratory frame) with velocity **v**, say, then it will be attracted to the wire in Figure 2.

Objectives 1 and 3   **SAQ 1**   Using the magnetic field as in Figure 2, verify the last statement.

For simplicity, let the test electron's velocity equal the current-carrying electrons' drift velocity, that is, let us take **v** = **u**. We may just as well view the experiment from some other inertial frame. (We assume the laboratory frame to be inertial, that is to say not rotating or accelerating. This is a fair assumption.) For convenience, let us take our second frame to move with the test electron, and let us denote the laboratory frame by S and the moving frame by S′. In S′, the electron is motionless but the wire moves to the left. The two situations are shown in Figures 3(a) and (b), where we have not indicated the magnetic field for simplicity.

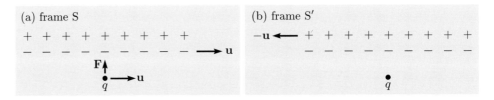

Figure 3   A test electron as seen in two frames.
(a)   In frame S, the test electron is moving to the right with velocity **u** and is attracted towards the wire.
(b)   In frame S′, the test electron is stationary. The lattice ions move to the left at velocity −**u**. Is the test electron attracted to the wire in frame S′?

We have seen that, in frame S, the electron is attracted to the wire. We shall now make the very reasonable assumption that an experimenter at rest in S′ will agree that the stationary test electron is attracted to the moving wire. If this were not the case, one observer might see the electron move closer and closer to the wire until it collided with it, while the other observer would see no such collision. Fortunately, such strange disagreements never occur in Nature!

Thus far, we have assumed only that Equation 1 applies to the specific frame S in which the wire is stationary. Let us now assume it is a law of physics to which the principle of relativity, originally enunciated by Galileo (see Unit 2, Section 5.4) applies. In other words, since all inertial frames are equivalent, we shall assume that Equation 1 takes the same form in all inertial frames of reference. That is to say, if **E′**, **B′**, **v′** and **F′** are the electric field, magnetic field, particle velocity and force as measured in frame S′, we assume

$$\mathbf{F} = q(\mathbf{E} + \mathbf{v} \times \mathbf{B}) \qquad \text{in S} \tag{2a}$$

$$\mathbf{F'} = q(\mathbf{E'} + \mathbf{v'} \times \mathbf{B'}) \qquad \text{in S'} \tag{2b}$$

Objective 1   **SAQ 2**   What additional assumption has been made in writing down Equation 2b?

In Unit 4, we pointed out that this assumption (that the Lorentz force law takes the same form in all inertial frames) has non-trivial consequences. In fact, it leads away from the Newtonian view of the world, placing us firmly on the relativistic path.

Now, in frame S, Equation 2a applies. For the situation depicted in Figure 3(a), we know that $\mathbf{E} = 0$ and the direction of $\mathbf{B}$ is such that the negatively charged test electron is attracted to the wire. (In the case of Figure 3(a), we have chosen that $\mathbf{v}$ equals $\mathbf{u}$.) But in frame S′, the test electron is stationary ($\mathbf{v}' = 0$). Thus the magnetic term $\mathbf{v}' \times \mathbf{B}'$ vanishes here. Since we are assuming that Equation 2b holds, by the principle of relativity, then we must conclude that *in S′ the electric field is not zero.*

Objectives 1 and 3   **SAQ 3**   What must be the direction of $\mathbf{E}'$ if $\mathbf{F}'$ acts in the same direction as $\mathbf{F}$?

Let us consider possible sources of the electric field $\mathbf{E}'$ in frame S′. You will recall that in Unit 4, Section 4.1, it was mentioned that it is known experimentally that time-varying magnetic fields create time-varying electric fields, and vice versa. However, in our experiment we have only a *steady state* situation: the wire is effectively infinitely long and all velocities are uniform. Thus the only possible source of $\mathbf{E}'$ is electrostatic (i.e. Coulombic). *In S, no electric forces act on q* and thus (in S) the wire is electrostatically neutral with equal numbers of positive ions and negative electrons per unit length. *In S′, however, there is a non-zero electrostatic field,* $\mathbf{E}'$.

Our assumptions have led us to conclude that in S′ the number of positive ions per unit length exceeds the number of electrons per unit length.

Objectives 1 and 3   **SAQ 4**   Prove the last statement.

Objectives 1 and 3   **SAQ 5**   Can you suggest a simple improvement to Figure 3(b), to make the Figure more nearly physically correct?

To repeat: Our series of not unreasonable assumptions has led us to the somewhat unsettling conclusion that a long wire carrying a steady current is electrically neutral in the laboratory system but carries a net charge per unit length when viewed from a moving inertial frame. This appears to conflict with one of the most basic laws of Nature (which is contained incidentally in the set of Maxwell equations), the law of invariance of charge, for have we not apparently 'created' an excess of positive over negative charge by going to frame S′? Indeed, in writing Equation 2b, we assumed charge invariance by setting q' equal to q. So isn't our argument inconsistent?

The answer to both questions can be 'no' provided we retreat one step. Let us continue to assume the law of the invariance of charge. Consider again the situation from the laboratory frame, S. Choose a convenient section, of length l, along the wire (you might think of placing two marks on the wire separated by a distance l). This section will encompass a certain discrete number of immobile positive ions, N, where $N = \lambda l$, λ being the number of ions per unit length. This is shown in Figure 4(a).

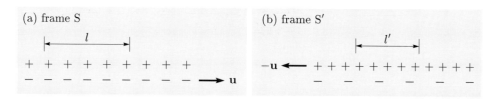

Figure 4   Distribution of ions in a wire.
(a) Length l contains a fixed number N of positive ions in frame S.
(b) The same number N of positive ions is contained in length l' in frame S′.

Now observe the same wire, and in particular the chosen section, from frame S′. We have agreed that in S′ the wire has an overall net positive charge per unit length. This is consistent with an *increase* in the number of positive ions *per unit length*. (One can also argue that there would be a *decrease* in the number of electrons per unit length, thereby enhancing the effect; see SAQ 6.) Thus by assuming that the law of charge invariance holds, we must insist that there are exactly N positive ions in the chosen section as observed in frame S′. With all our assumptions, then, the only possible conclusion is that the chosen section as observed in S′ is *shorter*; if this were so, then the number of positive ions per unit length (in S′) would have *increased*. For this reason, in Figure 4(b), we have indicated the length of the section as observed in S′ by l'.

To summarize the argument symbolically, we have concluded that if

$$N = \lambda l \quad \text{(in S)} \qquad \text{and} \qquad N' = \lambda'l' \quad \text{(in S′)},$$

then charge conservation implies that $N = N'$. Thus, $\lambda l = \lambda'l'$, or

$$\frac{l'}{l} = \frac{\lambda}{\lambda'}$$

so that an increase in the number of positive ions per unit length in frame S′ as compared with frame S (i.e. $\lambda' > \lambda$) implies that $l' < l$.

In brief, we may say that our series of assumptions has led us to conclude *that a given measuring rod appears to have different lengths when seen by different moving inertial observers.* As we shall see, according to Einstein's world-view this is indeed the case.

Objectives 1 and 3   SAQ 6   Can you give a simple argument to show that the number of electrons per unit length as observed in S′ might be *less* than the number per unit length in S?

This is as far as we wish to take the argument at this point. That the value of a measured length depends on the observer clearly runs counter to the Newtonian world-view, especially Assumption A10 of Block 1. That length is relative is, however, correct according to the new (1905) Einstein world-view. In Unit 6, when we have replaced the Galilean coordinate transformation (Unit 1, Section 5.3) by the new Lorentz transformation appropriate to the new Einstein world-view, we shall reconsider our thought-experiment. The *apparent* creation of charge in going from neutrality in S to a net positive charge per unit length in S' need not worry us, since current carrying wires *never* extend infinitely but always form closed loops; it will be shown that the net overall charge in a *closed loop is* zero for S', as it is for S.

Objectives 1 and 3   SAQ 7   Make a list of all the significant assumptions made in this section.

3 The principle of relativity

3.1 Newtonian mechanics and Maxwell electromagnetism

As you learned in Unit 2, the old Newtonian mechanics ($\mathbf{F} = m\mathbf{a}$) is invariant (unchanged in form) when going from one inertial frame, S, to any other inertial frame, S$'$; we would have $\mathbf{a}' = \mathbf{a}$ (Unit 1, Section 5.3), $m' = m$ and $\mathbf{F}' = \mathbf{F}$. In the Newtonian world-view, time is absolute and lengths are universal.

You have now seen, however, that by applying the principle of relativity to certain experimentally based facts it is possible to argue that lengths are *relative*. It seems then that we have touched on a fundamental disagreement between the new electromagnetism and the old Newtonian mechanics.

Historically, this disagreement centred primarily around the 'problem of light'.

You will recall that Unit 4 describes how Maxwell was able to deduce from his theory an equation, called the wave equation, which predicted that disturbances in electric and magnetic fields propagate through free space with no reference whatsoever to the original source of radiation. In setting up this equation, no medium for such wave propagation is introduced, unlike theories of acoustic- or water-wave propagation which are based on Newton's laws and *do* require the presence of a medium for their propagation. Even more strange (at the time) was the prediction of Maxwell's wave equation that his electromagnetic disturbances ought to move in free space with a speed

$$c = \frac{1}{\sqrt{\varepsilon_0 \mu_0}} = 2.998 \times 10^8 \,\mathrm{m\,s^{-1}} \tag{3}$$

(see Unit 4). The feature of this prediction that really bothered physicists in the second half of the nineteenth century was the fact that it makes no reference to the frame in which the electromagnetic disturbances are observed, since ε_0 and μ_0 are numbers easily measured in the laboratory without any reference to the propagation of waves. Remember that in Newtonian mechanics, the velocity of a physical disturbance depends on the inertial frame from which it is observed.

As a way out of this difficulty, some people suggested that Maxwell's wave equation did not apply in all inertial frames, but only in one select frame. This revives the old notion of an *absolute* reference frame devised by Newton for religious and philosophical reasons. You will recall from the Appendix in Unit 2, that Newton's attempts to demonstrate the existence of an absolute frame were in good measure thwarted by the invariance of his equation $\mathbf{F} = m\mathbf{a}$ under the Galilean transformation (Unit 1, Section 5.3) connecting coordinates between two such frames. Here, however, many hoped that one might be able to distinguish a 'select' frame.

It was natural, furthermore, to suppose that such an absolute reference frame might be the frame that was at rest with respect to a new kind of all-pervading medium with the rather peculiar mechanical property that it in no way hindered the motions of all other physical objects through it but provided the substance necessary (in the minds of many nineteenth century physicists) for the propagation of electromagnetic waves.

The first accurate measurements of the speed of visible light by Foucault in 1862 and Michelson in 1879 gave values very close to that predicted by Equation 3. Thus it was natural to identify visible light as an example of electromagnetic radiation. This was one reason why the supposed all-pervading medium for electromagnetic wave propagation was known as the 'luminiferous ether'.

Objectives 2 and 3   SAQ 8   Let us denote the speed of light through the luminiferous ether by c. Suppose a laboratory moves with some velocity **v** relative to the ether.

(a) If the laboratory is assumed to move entirely unhindered through the ether, what is the apparent speed of light beams directed parallel and antiparallel to **v**?

(b) The Earth moves about the Sun at a speed of about $30\,\mathrm{km\,s^{-1}}$. What is the maximum variation in the apparent laboratory speed of light over a 12 month period ?

In 1881, the American physicist Michelson (Figure 5) performed a subtle experiment with visible light to measure the velocity of his Earth-bound laboratory through the luminiferous ether. In simple terms, Michelson devised an apparatus called an interferometer with which he could compare the speed of propagation of light in two perpendicular directions; he might hope thereby to establish experimentally the existence of an 'ether wind' relative to his laboratory, due, among other things, to the Earth's motion about the Sun, which might advance or retard the propagation of light according to direction. The details of this elegant experiment are given in Appendix A and are discussed in a video band associated with this material. The content of Appendix A is examinable material, but has been left out of the main text for the sake of continuity.

We suggest that you read Appendix A before proceeding to Section 3.2.

Figure 5   A. A. Michelson

Figure 6   W. W. Morley

The result of Michelson's experiment surprised him; he found that the Earth's motion about the Sun through the supposed ether had *no effect* on the measured velocity of light in his laboratory. The accuracy of his experimental result was open to some doubt, however, and he was urged by Lord Rayleigh to improve his measurements. Michelson carried out a more accurate version of his experiment with the help of W. W. Morley (Figure 6), and confirmed the previous null result. Their apparatus is shown in Figures 16 and 17, in Appendix A. This experiment, known as the Michelson–Morley experiment, has since been repeated in more refined versions as lasers and other new devices of experimental physics have appeared.

3.2   Attempts to modify electromagnetism before Einstein

The unexpected and definite null result of the Michelson–Morley experiment posed severe problems, but a number of ingenious *ad hoc* mechanisms were soon proposed to get out of the difficulties and preserve the ether concept more-or-less intact. One of these was the 'ether-drag' hypothesis, wherein it was suggested that the ether was dragged along by all solid bodies such as the Earth, thus ensuring that both beams of light in the Michelson interferometer would travel at the same speed. This explanation was quite untenable, however, as certain stellar aberration observations are quite inexplicable if there were an ether which moved with the Earth.

An alternative suggestion was put forward independently by G. F. Fitzgerald and H. A. Lorentz in 1892. This was that all bodies are contracted by their motion through the ether, the contraction being in the direction of motion and of a magnitude just sufficient to equalize the times taken by the two beams in the interferometer. It can readily be shown that contraction by the factor $\sqrt{1 - (v^2/c^2)}$ would explain the null result of the Michelson–Morley experiment, but not, incidentally, the null result of the Kennedy and Thorndike modification of this experiment (to be described later in Appendix B4).

The Fitzgerald–Lorentz hypothesis was essentially *ad hoc*. Nevertheless, it did happen to hit on one feature which was to emerge from Einstein's treatment of the problem (see Section 3.1 of Unit 6). It gave no hint, however, of the further consequences that were to arise from Einstein's theory, particularly those affecting properties of time.

The modern view of the Michelson–Morley result is simply to accept it as a clear demonstration of the *isotropy of space* with respect to light propagation. (This will be considered at more length in Appendix B3.)

In the next section, we begin to discuss Einstein's world-view and in particular his approach to light.

3.3   Einstein reasserts the principle of relativity

In a nutshell, Einstein's 1905 paper extended the Galilean principle of relativity to assert that all inertial frames are completely equivalent for *all* physical phenomena. In particular, its scope was to be extended beyond the domain of particle dynamics to include the phenomena described by Maxwell's electrodynamics. Motivated by thought-experiments not unlike that in Section 2 of this Unit, Einstein argued that the mathematical expressions obtained by Maxwell to describe the new electrodynamics apply *as they stand* in any inertial frame (we made this assumption in writing Equations 2a and 2b).

One result of Maxwell's theory is the wave equation, which predicts that electromagnetic waves, and in particular visible light, propagate in free space with speed c. Einstein's extension of the principle of relativity meant that this same speed would apply in all inertial frames, independent of direction and independent of whatever motion the source might have.

The 1905 paper condenses these considerations into two postulates:

Postulate SR1   The principle of relativity

The laws of physics take the same form in all inertial frames [that is, the laws of physics do not distinguish one inertial frame from another].

Postulate SR2   The principle of 'constancy' of the speed of light

In particular, the speed of light in free space has the same value in all inertial systems.

These two postulates are the foundation of what is now known as Einstein's special theory of relativity. It is 'special' in that it does not take into account the effects of gravity. To do this, Einstein had later to extend his treatment. His general theory of relativity is introduced in Block 3. There we shall find, for example, that the concept of 'inertial frame' must be reinterpreted. Because the presence of matter affects the nature of space and time in its vicinity, 'inertial frame' in Block 3 loses the global nature attributed to it in Blocks 1 and 2 and becomes local in character.

The crucial implication of Postulate SR2 is the constancy of the speed of a given beam of light as measured in any two or more inertial frames of reference, even though the frames of reference may be in relative motion. Postulate SR2 has an even stronger meaning than this, however, for it emerges from experiment (or from Maxwell's equations, which after all merely summarize all our experimental observations) that the speed of light (or more generally, any electromagnetic signal whatsoever) in free space has the same value for all time, independent of its frequency or the nature and motion of its source, independent of the direction of propagation or the time and place of measurement, and independent of the particular inertial frame with respect to which this speed is measured.

Postulate SR2 appears at first sight to be irreconcilable with SR1. The difficulty, of course, is that the Galilean coordinate transformation equations require that two observers in relative motion will generally obtain different results when they measure the speed of any particle, including 'particles' of light (Equation 23 of Unit 1). On the other hand, it is verified by experiment (as we shall see in Appendix B) that the speed of light is the same for two such observers. Clearly, *Galilean transformations must be the wrong transformation between inertial frames.*

This would be a good point at which to study Appendix B.

The present-day evidence for the constancy of the speed of light is so strong that the invalidity of the Galilean transformations of space and time is beyond doubt. But at the end of the nineteenth century, this was still far from clear. In fact, it was quite unsuspected until Einstein brought the matter sharply into focus by pointing to the inadequacies of the Newtonian concept of time.

4   Time in special relativity

4.1   Introduction

Implicit in the Newtonian concept of absolute time is the assumption that a definite time order can be assigned to events independent of the position of these events in a given inertial frame of reference (Assumption A8 of Block 1) and, more significantly, independent of the particular inertial frame of the observer (Assumption A9 of Block 1). The truth of these assumptions had been virtually unquestioned since the time of Newton, until Einstein began to ponder on just how in practice one might go about assigning times to events.

In considering this question, Einstein first of all emphasized the importance of the concept of *simultaneity*. We quote from a translation by J. A. Barth of Einstein's 1905 *Annalen der Physik* paper:

> If we wish to describe the *motion* of a material point, we give the values of its coordinates as functions of the time. Now we must bear carefully in mind that a mathematical description of this kind has no physical meaning unless we are quite clear as to what we understand by 'time'. We have to take into account that all our judgments in which time plays a part are always judgments of *simultaneous events*. If, for instance, I say, 'That train arrives here at 7 o'clock,' I mean something like this: 'The pointing of the small hand of my watch to 7 and the arrival of the train are simultaneous events.'

Thus *the assigning of times to events involves judgements of simultaneity*, and certainly we could achieve an absolute Newtonian time-scale if all observers, independent of their position and velocity, agreed whether two events (for example the train arriving at the station and the pointing of the hour hand of a given clock to 7) are *simultaneous*.

4.2   A thought-problem: implications of assigning times

To come to grips with the problem of assigning times to events separated in space, let us consider a simple thought-experiment in the same spirit as the experiment of Section 2. Recall that in that experiment different inertial observers encountered interesting complications when attempting to assign spatial lengths to the distance separating two points in space. To present the problem, we shall continue to use the somewhat informal style of the discussion in Section 2. The methods used here will be seen to be justified later in the Block.

The invariance of the speed of light has far-reaching implications in the new Einsteinian world-view. To see this, consider the experiment depicted in Figure 7. A rod lies at rest in a frame called its 'rest frame', which we denote by S (Figure 7a). Since we assume space is homogeneous, there is no difficulty in supposing that we can locate the midpoint of the rod. We imagine exploding a flashbulb there. Subsequently, light will move out in all directions *at speed c*. We shall concentrate on the two pulses moving along the rod in opposite directions. Clearly, these two pulses will arrive at opposite ends of the rod at *the same time*.

Now imagine the same experiment as observed from an inertial frame moving, to the right say, with velocity **V**. In this frame, denoted by S′, the rod moves to the left with velocity −**V** (Figure 7b). It is an interesting,

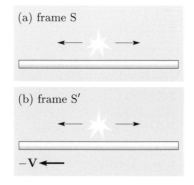

Figure 7   Two pulses moving outwards from the centre of a rod.

(a)   In S, the pulses arrive simultaneously at the ends of the rod.

(b)   In S′, the rod moves to the left at velocity −**V**, but the pulses still move at speed c.

and to some people even an unsettling consequence of postulate SR2 that the two pulses move along the rod in S′ with the *same* speed c as they do in S. Whether one likes it or not, this fact has been experimentally verified and so must be accepted. All we can do is to work through the logical outcome of the postulates. We ask whether or not the pulses arrive simultaneously at the ends of the rod as observed in S′. Before proceeding, you might want to guess which pulse will arrive first (in S′) — if indeed they do not arrive simultaneously.

Because the rod moves to the left in S′, the right-hand end of the rod 'runs up' to meet its pulse, whereas the left-hand end 'runs away from' its pulse. Thus, in S′, the right-hand pulse arrives *before* the left-hand pulse.

Remember, we are viewing the *same* process from another frame — and in that frame, S, the light reached each end simultaneously. In Einstein's world-view, simultaneity is no longer necessarily absolute.

We shall reconsider this thought-problem in more detail in Section 7. There we shall conclude that two events occurring simultaneously at the *same* point in space according to one inertial observer are observed by all inertial observers to be simultaneous. *Simultaneity, however, is relative to the observer when the two events are separated in space.*

Objective 5   SAQ 9   By imagining the length of the rod to approach zero, can you give a simple argument to show that two events not separated in space are simultaneous for all inertial observers if they are simultaneous for any one such observer?

4.3   Synchronizing clocks in an inertial frame

From the argument of the preceding section, it is clear that a careful rethinking of the assignment of times to events is needed. In this section, we begin by asking how one might attach times to events at various places in a single inertial reference frame.

Assigning times to events in the *immediate vicinity* of a clock is quite simple and unambiguous as the answer to SAQ 9 shows. However, in describing the motion of particles, etc., we need to assign meaningful times to events at many different places. What we might use for this, in principle at least, is *many* clocks. If all of them are *synchronized* with each other, then we can imagine placing a synchronized clock, at rest, at the location of every event to which we may wish to assign a time. This may require 'seeding' entire regions of space with such synchronized clocks. In this way, the time of every event can be read off a stationary clock, *located at the event*, by a data taker located *at that clock*. But what exactly do we mean by synchronized clocks, and how do we achieve this synchronization?

The following three requirements must be satisfied by the synchronization process:

1 In any one inertial reference frame, a *unique* time is to be assigned to any given event occurring at any point (x^1, x^2, x^3).

2 The process of synchronizing the array of clocks in any one inertial frame is to follow the spirit of Postulate SR1. *The synchronizing process is to be the same for all inertial observers.* In other words, although observers in different inertial frames might disagree whether a particular set of clocks is synchronized, there should be just one procedure that applies in each inertial frame.

3 When synchronized, any two clocks from an array of synchronized clocks are synchronized with each other; thus if A, B, and C are any three clocks, we may say that if pairs A, B and A, C are synchronized, then the pair B, C is synchronized.

To achieve synchronization in the spirit of these requirements, one possibility might be to adjust all the clocks to read the same time when situated at one place, the origin say, and then to move them to the desired positions. Naturally, we would have to choose some kind of clock that isn't affected by being moved about. This would rule out various pendulum clocks, water clocks, and Big Ben. Some kind of spring-driven clock or atomic clock would be suitable. But even with these (and indeed with any others) it is necessary to move them *infinitesimally slowly* to their assigned positions in space. This is necessary owing to fundamental properties of spacetime. For, according to the famous twin 'paradox' of special relativity (to be discussed in Units 6 and 7), if one of two clocks which initially read the same time at one point in space is separated from its pair and then returned at finite speeds, the two clocks will then *not* record the same time. This will be considered in Section 4.2 of Unit 6 and Section 5.3 of Unit 7.

A better alternative would be to synchronize all the clocks *after* moving them to their various fixed positions. If we could transmit information (in this case the positions of the hands of the various clocks) with infinite speed, we could easily adjust the readings on the various clocks to agree with one master clock located at, say, the origin of coordinates. But we must pay heed to the realities of Nature: signals can only be transmitted at finite speed. Einstein suggested we use light signals to synchronize our clocks.

Objectives 4 and 6   SAQ 10   Give an argument that light signals, rather than some other sort of signal (sound, perhaps), should be used to synchronize our clocks.

Objectives 4 and 6   SAQ 11   The speed of light had been measured long before the birth of Einstein. This, of course, involved the measurement of time intervals. Can you say why physicists were able to measure the speed of light correctly using the Newtonian concept of time?

How can we synchronize the clocks using the light signals? One might naively suggest a method whereby an observer at some fixed point (the origin of the coordinates, say) actually *sees* them as all reading the same time. To put this another way, let each clock emit a flash of light when its hands point to 12 o'clock, say; then the clocks would be 'synchronous' if all these light signals reached the observer simultaneously. Of course, the trouble with this method of synchronization is that it depends on the *position of the observer*. Clearly, an observer at a different position will not see the clocks synchronous by this definition, as his distances from the various clocks, and therefore the 'delay times' due to the finite speed of propagation, will be different from those of the first observer. This method thus fails as the next SAQ makes clear.

Objectives 4 and 6   SAQ 12   Show that this method of 'synchronizing' clocks would not satisfy the first requirement. Would the method satisfy the first requirement if the speed of light were infinite?

Einstein adopted the following procedure for synchronizing clocks, which we cannot explain more clearly than he does in his 1905 paper:

> If at the point A of space there is a [stationary] clock, an observer at A can determine the time values of events in the immediate proximity of A by finding the positions of the hands which are simultaneous with these events. If there is at the point B of space another [stationary] clock in all respects resembling the one at A, it is possible for an observer at B to determine the time values of events in the immediate neighbourhood of B. But it is not possible without further assumption to compare, in respect of time, an event at A with an event at B. We have so far defined only an 'A time' and a 'B time'. We have not defined a common 'time' for A and B. The latter time can now be defined in establishing *by definition* that the 'time' required by light to travel from A to B equals the 'time' it requires to travel from B to A. Let a ray of light start at the 'A time' t_A from A towards B, let it at the 'B time' t_B be reflected at B in the direction of A, and arrive again at A at the 'A time' t'_A.

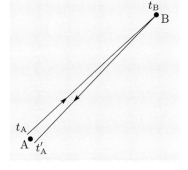

Figure 8   Place the hands of clock B at the setting $t_B = \frac{1}{2}(t'_A + t_A)$, and start clock B at the instant the light beam reaches B. Clocks A and B are then sychronized.

In accordance with this definition, clock B is synchronized with clock A if

$$(t_B - t_A) = (t'_A - t_B) \tag{4a}$$

i.e. if the clock at B is set to read such that

$$t_B = \frac{1}{2}(t'_A + t_A). \tag{4b}$$

This is depicted in Figure 8. Equation 4b, then, is the condition that clock B is synchronized with clock A. For later use, note that

$$t_B - t_A = AB/c. \tag{4c}$$

This procedure is equivalent to the alternative procedure of placing a light source *exactly midway* between A and B, setting both the clocks to the same reading $(t_1 = t_2)$, and starting each clock at the instant a signal resulting from a single flash of light from the source reaches the clock. This latter procedure is depicted in Figure 9.

Objective 6   SAQ 13   What basic property of light ensures that the two synchronization procedures illustrated in Figure 8 and 9 are equivalent?

In each case, the method assumes, by *definition*, that the speed of light is the same in both directions, backwards and forwards along a given path. It follows that if clock B has been made synchronous with clock A by the procedure of Figure 8, then clock A is automatically synchronous with clock B. This symmetry is quite obvious if the procedure of Figure 9 is used.

Objective 6   SAQ 14   Show that if clock B is made synchronous with clock A by the procedure of Figure 8, then a subsequent flash of light emitted from B to A and back to B would demonstrate that clock A is synchronous with clock B.

Figure 9 Adjust the clocks so that $t_1 = t_2 = t$, say, at the instant that light signals are received at A and B respectively. Clocks A and B are then synchronized. (O is the mid-point of AB.)

By using Equation 4, we can synchronize any clock B with some standard clock A. We can imagine successively carrying out this synchronization procedure with all our clocks at their various positions in our reference frame, setting them all to be synchronized with clock A. But does synchronization defined in this way have any meaning independent of our arbitrary choice of the 'standard' clock A? Can we speak of all our clocks, synchronous with A, as being *synchronous with each other*, so that no particular clock is favoured as a standard? In other words, can we show that this method of synchronization satisfies the third requirement? We can, provided the speed of light is the *same in all directions* in our frame of reference, as Einstein assumed as a fundamental hypothesis and as has been experimentally verified.

Objective 6   SAQ 15   Taking the speed of light as the same in all directions in our frame of reference, and assuming stationary clocks at points A, B and C, show that if clocks A and B are synchronous, and clocks A and C are synchronous, then clocks B and C are also synchronous with each other. [*Hint*: Use Equation (4c) and others like it.]

It appears, then, that we have a method of synchronizing clocks that meets all three requirements. We may choose *any one* of our array of clocks and then synchronize all the other clocks to our chosen clock. The method then ensures that any pair of clocks will be synchronized. This is spelled out in the following SAQ.

Objective 6   SAQ 16   Suppose we wish to synchronize a three-dimensional array of clocks spread out uniformly in three-dimensional space. Let us choose to do this by exploding a flashbulb at the origin, say. We arrange the clocks beforehand so that the ensuing spherical pulse of light sets the clocks going from their appropriately synchronized time. Since the speed of light is independent of direction, all clocks are located a distance R from the origin will be set to start at the same reading, t_R. Give an expression for this time setting if the flash is set off at time zero on the clock at the origin.

In summary, we see that by synchronizing clocks according to Equation 4, we have defined a measure of time for all points within our frame of reference by means of a set of clocks at rest in our frame of reference. This definition relies on the observed fact that the speed of light is the same in all directions, which is implied by Postulate SR2. This postulate also provides that the speed of light is the same when observed from different frames in uniform translatory motion relative to one another, so that the *same* prescription, Equation 4, is equally valid, without prejudice, for synchronizing clocks at rest within *any* particular inertial frame of reference. In other words, this method of synchronization satisfies the first requirement as well as the second.

By way of further summary, consider the following set of questions and answers:

QUESTIONS   An observer has synchronized the clocks in his frame according to the above procedure. Standing at one point in the frame, he sees all the (fixed) clocks running at the same rate, and he sees with his eyes that the hands on all clocks *equidistant* from him are pointing to the same time. How do the times he sees (with his eyes) shown on the clocks vary with their distances from him in any direction? Put another way, if at any given instant he were to take a photograph of all the clocks, how would the readings of all the clocks, as seen on the photograph, vary with distance? If he had made his observations from any other point in the frame, would they have differed in any way?

Now imagine that *after* the clocks were synchronized, the speed of light was reduced. What would he now see? Would the clocks still be synchronized? What would he see if the speed of light were increased? What would he see if the speed of light were made infinitely large (an entirely non-physical possibility!)?

Finally, *imagine* that the observer can attain an overall view of the clocks by moving from point to point at infinitely high speed. What would he now see?

ANSWERS   Because of the finite speed of propagation of light (from the clocks to the observer), the images formed on the retina of his eye or on his photograph show the readings of distant clocks as they *were* at an earlier time. The time difference between the reading he sees on a distant clock and the one he sees on a clock very close to him is accounted for by the time taken by light to propagate from the distant clock to his retina, carrying with it the image of the clock. This propagation time will be proportional to the distance of the clock from the observer, so he sees distant clocks as showing times earlier than his local clock by amounts that increase in direct proportion to his distance from these clocks. Because the speed of light is the same in all directions this effect does not depend on direction. Because the speed of light is the same everywhere and because all the clocks are mutually synchronous (see SAQ 16), when he moves to any other point he will again see all clocks equidistant from him as showing the same time, with this time being further in his past for clocks further away from him.

Now, if the speed of light is imagined to be *reduced* (*after* the original synchronization has been carried out using 'real' light) he would see the same sort of thing, but clocks at a certain distance from him would now be reading times even further into his past than they did before. If the speed of light were *increased* after the initial synchronization, distant clocks would again show readings in his past, but the time differences would now be smaller, corresponding to the smaller times of propagation of this 'fast light' from the clocks to the observer. For 'infinitely fast light' these propagation times would all vanish, and he would see every clock indicating the *same* time. So an alteration to the speed of light would change the appearance of the clocks as seen by the observer; it would not, however, change the synchronization of the clocks, for this is based on the general invariance of the speed of light with respect to observer and direction.

He could reduce the delay caused by the propagation time of real light if, instead of waiting for the light to reach him from a given clock, he moved towards the location of that clock. If he could do this with infinite speed, the effect would be the same as seeing from his original vantage-point, images of the clock as transmitted by 'infinitely fast light'; therefore by moving from point to point at infinite speed, he would obtain an overview of all the clocks indicating the *same* time. (We are assuming, here, that we have a short-sighted observer of zero mass who can only see the clocks in his immediate vicinity!)

5 Taking stock

We began this Unit (Section 2) with an argument that revealed flaws in the Newtonian world-view: In particular, it appeared that if the Lorentz force law is valid in all inertial frames, then different observers would find different values for observed lengths. There seemed to be a contradiction between the old Newtonian mechanics and the new electrodynamics which deals with effects outside the Newtonian domain.

Maxwell's wave equation could be interpreted (Section 3.1) as implying that electromagnetic disturbances, such as light pulses, propagate with speed c only in a reference frame fixed with respect to a medium called the luminiferous ether. But there was another interpretation: according to this, light and other electromagnetic disturbances do *not* require a medium in which to propagate. Maxwell's equations are valid for *all* inertial frames, and the speed of light is the same for *all* inertial observers.

The natural interpretation of the null result of Michelson and Morley (Section 3.1 and Appendix A) was a refutation of the ether hypothesis. Several *ad hoc* hypotheses, especially the Lorentz–Fitzgerald contraction hypothesis (Section 3.2), were proposed to try and salvage the ether concept. But the confusion came to an end in 1905 with Einstein's famous paper (Section 3.3). In this, he asserted that the principle of relativity applies to all natural phenomena (Postulate SR1) and that the speed of light is constant irrespective of direction of propagation, region of space and motion of source, and is invariant for all inertial observers (Postulate SR2). In effect, this asserted that Maxwell's equations are laws to which SR1 applies.

To underline the fact that the postulates of special relativity do lead to the prediction of physical effects at odds with our everyday perception, we considered a thought-experiment which showed that time as well as length is relative; in Section 4.2, we argued that two events occurring at different points in space, though simultaneous for one inertial observer, will not be simultaneous for other inertial observers moving with respect to the first.

The question was then raised as to whether meaningful times could be assigned to events in any single inertial frame. Section 4.3 described a process based on the constancy of the speed of light whereby a field of clocks lying at rest in any inertial frame can indeed be synchronized so that meaningful times can be assigned to events at all spatial points.

6 Observers and world-lines in special relativity

Figure 10   The world-line of a particle

In Section 4.3, we saw how to assign unambiguous times to events in any inertial frame. We now consider the following question: 'A particle moves along the x^1-axis in an inertial frame. How does one measure its velocity as a function of time?' Let us consider the question in the context of special relativity.

We might try to draw the world-line of the particle in the usual way as was done in Unit 1 — see Figure 10. But this kind of picture is built on the idea of time being absolute. Is it still valid to draw a world-line in this way? If time is no longer absolute, how can we draw such a set of axes purporting to give the precise time (or times) at which the particle was located at any position x^1? This has been discussed in Section 4.3, where it is shown that we can think of a whole field of data takers located at different positions all along the x^1-axis, each provided with *identical, synchronized* clocks. Each data taker is given the job of recording the time (or times if the particle retraces its steps) on *his* clock when the particle passes *his* position x^1. We may then visualize all observers meeting together at the origin $x^1 = 0$, say, and reporting their data. Figure 10 then applies, with t being the time recorded by the relevant observer on his synchronized clock.

We shall use spacetime diagrams like Figure 10 for the rest of the Block and, indeed, for the rest of the Course. For convenience, let us agree that in such spacetime diagrams, we shall use the variable ct for the time coordinate, where $c \sim 2.99 \times 10^8\,\mathrm{m\,s^{-1}}$ is the speed of light and t is the time in seconds. (Of course, ct has dimensions of length.) Thus, consider an experiment where an apparatus situated at the point $x^1 = 0$ emits a short, sharp, pulse of light out in all directions at time $t = 0$. Ignoring the x^2 and x^3 axes, we may visualize the world-line of the pulse as in Figure 11. Because we are using ct, and not just t, the slope of the trajectories is $+1$ or -1, that is, the world-lines of the light pulse exactly bisect the first and second quadrants.

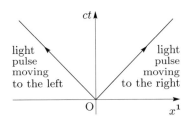

Figure 11   The world-lines of light pulses

The equation for the world-lines in this case is simple, for according to Postulate SR2 (Section 3.3) the *speed* of light will surely be c. Thus, since the pulse will spread out in all directions from the point $x^1 = 0$, we can say that the world-line to the right will be given by $x^1 = ct$ while that to the left will be $x^1 = -ct$.

Much of the argument in this and the next two units involves the use of spacetime diagrams; the following four SAQs are intended to give you a little practice in using them. You should make sure that you thoroughly understand the answers to these SAQs.

Objective 7   **SAQ 17**   Consider an observer at rest at some point $x^1 = a$ in an inertial frame. Draw his world-line in the (ct, x^1) coordinate system for times $t > 0$.

Objective 7   **SAQ 18**   For the experiment with the light pulse just described in the text, why was the world-line of the pulse not drawn as in Figure 12?

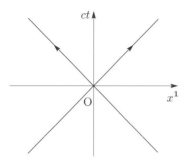

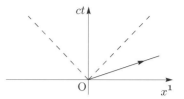

Figure 13   For SAQ 19;
the dashed lines are at 45° to
the axes.

Figure 12   For SAQ 18

Objective 7   SAQ 19   As you will see in Unit 7, the theory of special relativity leads to the conclusion that no material body can be accelerated to speeds equal to (or greater than) c. Suppose a particle moves uniformly to the right along x^1 starting at $x^1 = 0$ when $t = 0$. Does Figure 13 show a possible world-line? If not, why not?

Objective 7   SAQ 20   A pulse of light is emitted from $x^1 = 0$ to the right along the x^1-axis. A stationary mirror is located at some distance $x^1 = a$ to the right of the origin. Draw the world-line of the light pulse as it moves to the right and then is reflected back along x^1.

We have indicated several times that the term 'observer', as used in a technical sense, can be taken to mean a whole array or system of data takers (or observers in the usual sense) arranged at fixed points, each with an identical synchronized clock. This notion of the process of data-taking is perfectly valid in principle. But in actual physical measurements, one or two data takers must make all the observations of distant events. Operationally, then, cannot one imagine a *single* observer at the origin, say, measuring the correct time and position of distant events?

Yes, there is such a possibility. Let us consider the following thought-experiment: an event occurs at some point $x^1 = a$, to the right of the origin, at some time t. How can a single observer at $x^1 = 0$ assign the correct time to the event? We might think of the observer sending consecutive coded pulses spaced closely and uniformly in time as he measures time on *his* clock at *his* position. We could imagine some apparatus at $x^1 = a$ such that when the event occurs *there*, the particular coincident incoming light pulse received *there* is reflected back to the single observer at $x^1 = 0$. Since the pulses are coded somehow (for example by their colour), the experimenter at $x^1 = 0$ knows that he sent that particular pulse out at time t_{out}, say, on *his* clock. Suppose he receives the pulse back at time t_{back} on *his* clock. What time should be assigned to the event? According to Equation 4b in Section 4.3, we have $t_{\text{event}} = \frac{1}{2}(t_{\text{out}} + t_{\text{back}})$.

Objectives 6 and 7   SAQ 21   Sketch the world-line of the particular light pulse under discussion. Label t_{out}, t_{back}, t_{event}, and $x^1 = a$.

Objectives 6 and 7   SAQ 22   Suppose that the single observer at $x^1 = 0$ didn't know that the event occurred at $x^1 = a$. How could he calculate this from his knowledge of t_{out} and t_{back}?

Objective 6   SAQ 23   In a few sentences, give two equivalent operational definitions of time which could apply to any point in a chosen inertial frame.

Objectives 6 and 7   SAQ 24   In a few sentences, explain what we mean by 'observer' and 'spacetime diagram'.

Throughout much of the rest of this Block, we shall be considering only one-dimensional thought-experiments, that is, we shall primarily consider events that can be located in spacetime by the coordinates (ct, x^1). Since we shall have quite a bit of (straightforward) algebra, let us adopt the following conventions:

1 Henceforth in this Block, let us agree that whenever we are dealing with a one-dimensional thought-experiment, we shall often choose the x^1 coordinate and write x instead of x^1. However, when we must refer to three dimensions, we shall continue to write x^1, x^2 and x^3.

2 Furthermore, we have seen that it is feasible to think of a single observer, at the origin, say, who can ascertain the coordinates (t, x) — or (ct, x) if you prefer — of an event occurring at a distance. Therefore let us agree to identify the specific system of coordinates by the phrase 'an observer O'.

Henceforth in this Block, we shall use the symbols O, O$'$, O$''$, etc., to represent any observer in the new special-relativistic sense of *observing system*. In this way, we formally break with the old pre-Einstein picture with which we started this Unit, where we used frames S, S$'$, etc., before we were aware of the problem of assigning times to events.

With these conventions settled, let us now enquire how an observer O can determine the world-line of a particle moving in some way relative to him.

So far, we have considered how an observer O can assign coordinates to any single event $\mathscr{E}$. We shall consider, as a final example, how O would measure the path (world-line) of a particle moving along in an arbitrary way.

The continuous world-line of a particle represents a continuous series of events. For example, we have shown in Figure 14 two such adjacent events labelled $\mathscr{E}_A$ and $\mathscr{E}_B$. As calculated in the answer to SAQ 22, any single event can be assigned its coordinates (ct, x), where

$$(ct, x) = \left(\frac{c(t_{\text{out}} + t_{\text{back}})}{2}, \frac{c(t_{\text{back}} - t_{\text{out}})}{2} \right). \tag{5}$$

Thus we can assign coordinates (ct_A, x_A) and (ct_B, x_B) to $\mathscr{E}_A$ and $\mathscr{E}_B$.

To see this, imagine a small mirror fixed to the particle. The experimenter at $x = 0$ can then send two pulses out along the x-axis at two adjacent times t_1 and t_2 on his clock. These two pulses will be reflected back to him and their times of return, t_3 and t_4 on his clock, can be recorded. The whole process is depicted in Figure 14. The coordinates of event $\mathscr{E}_A$ are:

$$(ct_A, x_A) = \left(\frac{c(t_1 + t_3)}{2}, \frac{c(t_3 - t_1)}{2} \right).$$

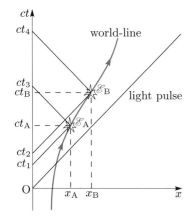

Figure 14   Two events, $\mathscr{E}_A$ and $\mathscr{E}_B$ on the world-line of a particle.

Objectives 6 and 7   SAQ 25   Give the coordinates of event $\mathscr{E}_B$ in Figure 14.

The velocity in the region near $\mathscr{E}_A$ and $\mathscr{E}_B$ along x is given approximately by $\Delta x/\Delta t$ where $\Delta x = x_B - x_A$ and $\Delta t = t_B - t_A$. By varying t_1 and t_2, the experimenter can make $(t_2 - t_1)$ infinitesimally small. From Figure 14 it is clear that when $(t_2 - t_1)$ becomes very small, so will $(t_B - t_A)$. Thus an increasingly accurate measurement of the particle's velocity at a point in spacetime will be given as

$$\lim_{t_2 - t_1 \rightsquigarrow 0} \left(\frac{\Delta x}{\Delta t} \right) = \lim_{t_2 - t_1 \rightsquigarrow 0} \left(\frac{x_B - x_A}{t_B - t_A} \right).$$

We began this section with a spacetime diagram in the spirit of the old Newtonian world-view but, by drawing on the various strands developed thus far in the Block, we have indicated that spacetime diagrams can be used precisely and unambiguously to represent events and world-lines from the new special relativistic world-view.

We would like to emphasize that, just as for the Newtonian world-view, each observing system will have its own spacetime diagram to represent events and world-lines. For instance, a particle lying at rest in one frame will be observed to move uniformly in any inertial frame moving relative to the first frame; the world-line in the first frame is a vertical line parallel to t-(or ct-) axis, while in the second frame the world-line is tilted.

We cannot overemphasize the usefulness of spacetime diagrams, especially in portraying the qualitative features of thought-experiments in special relativity. We shall make great use of spacetime diagrams throughout the rest of the Block.

event

A final point of terminology: the word 'event' has crept into the discussion and it will occur very often wherever relativity is the subject. In this context, it has a somewhat abstract meaning which departs a little from the everyday sense in which an event is something that takes place over a finite interval of time and a finite volume of space (think of 'events' at Glastonbury, Wembley, Covent Garden ...). An *event* in our sense is an occurrence at an *instant* of time and a *single point* in space. The essential feature of any event is therefore specified by its time t and its coordinates x^1, x^2, x^3. In Unit 7, we shall actually define 'event' in special relativity as a single point on a spacetime diagram, but if you look back over this last section you will see that, in effect, we have already been using it in a sense which is rather close to that.

7 The simultaneity problem revisited

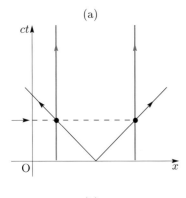

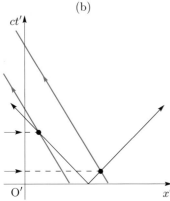

Figure 15   The arrival of two light pulses at the two ends of a rod.

(a) The events observed in O, the rest frame of the rod.

(b) When the same events are observed in frame O′, simultaneity is lost.

Using concepts developed thus far, in particular noting that each observing system has its own spacetime diagram, we can now think again about the experiment described in Section 4.2. This time we use spacetime diagrams — one for an observer in each frame, as in Figures 15(a) and 15(b).

In Figure 15(a) we show the experiment as construed by observer O. (O is the observing system, or observer, attached to the frame which we labelled S in Section 4.2. We often refer to observer O's frame as frame O, indicated with 'O' at the origin.) Since the rod is at rest, its ends lie at given constant values of x as time proceeds. Thus the rod's ends are represented by two vertical blue lines in Figure 15(a). The two light pulses, represented by black lines, move out to the left and right along the rod.

Since the speed of light (by postulate SR2) is the same to right and left, the two black lines intersect the x-axis at the same angle. The intersections of the blue lines with the black lines (i.e. the arrival of light at the ends of the rod) occur at the same time in O (i.e. at the same value of ct, as c is constant). In O, these are simultaneous events.

Figure 15(b) shows the spacetime diagram for system O′ with axes (x', ct'). (In Section 4.2, we labelled the frame of observer O′ by S′.) The two light pulses move out, by Postulate SR2, at the same speed c. Thus their trajectories intersect the x'-axis at the same angle, as represented by two black lines. In O′, however, the rod moves to the left uniformly. The spacetime paths of the two ends of the rod are therefore tilted to the left, as indicated by the two blue lines in Figure 15(b). Where the black lines intersect the blue lines, the two light pulses reach the ends of the rod. Figure 15(b) makes it clear, as anticipated, that in O′ the right-hand end of the rod is observed to encounter its pulse before the left-hand end receives its pulse. In frame O′, the two events defined by the arrival of the light signals at the ends of the rod are not simultaneous as they were in O.

Objective 5   **SAQ 26**   Repeat this analysis briefly for the case of an observer O′ whose frame moves to the *left* with respect to the rest frame O. Sketch the spacetime diagram analogous to Figure 15(b).

Objectives 4 and 5   **SAQ 27**   In a nutshell, what accounts for the loss of simultaneity between frames O and O′?

Summary

1 By considering the force exerted on a charged particle by a current-carrying wire as observed from two inertial frames, it was seen (in Section 2) that a magnetic force in one frame could be interpreted as an electric force in the other. Moreover, this thought-experiment indicated that there was reason to question the Newtonian assumption that length was an invariant quantity.

2 It was noted in Section 3.1 that Maxwell's electromagnetic theory led to a value for the velocity of light that was dependent on the measurable quantities ε_0 and μ_0. The question was raised of whether there was some special frame of reference to which this velocity referred, a frame that was perhaps defined by being at rest relative to a medium for the light waves (the 'luminiferous ether').

3 The Michelson–Morley experiment, described in Appendix A, demonstrated that the speed of light is independent of the direction of propagation. This observation was at variance with the hypothesis of a preferred frame of reference for light propagation.

4 The Michelson–Morley experiment could be explained in terms of a 'ballistic' theory of light. This hypothesis (discussed on the videotape), however, was discounted by observations on binary stars that showed that the speed of light does not depend on the motion of the source. This experiment, and others showing that the speed of light is independent of frequency, the time and position of the measurement, and of the particular inertial frame from which the measurement is made, are described in Appendix B.

5 Einstein's special theory of relativity is based on two postulates (described in Section 3.3);

> **Postulate SR1 The principle of relativity**
> The laws of physics take the same form in all inertial frames.
>
> **Postulate SR2 The principle of the constancy of the speed of light**
> In particular, the speed of light in free space has the same value in all inertial systems.

6 In Section 4.2, a thought-experiment was described which demonstrated that as a direct consequence of Postulate SR2, observers in relative motion will in general disagree about the simultaneity or otherwise of events separated by a distance. This conclusion cannot be reconciled with the Newtonian concept of absolute time.

7 The above conclusion pointed to the need for a careful re-examination of the fundamental concept of time. As a first step, we established a procedure for assigning a time to an event occurring at any point in space — a time appropriate to a particular inertial reference frame. There are two slightly different variants of this procedure: one involves the mutual exchange of light pulses between equivalent observers distributed throughout space, each possessing their own clock (this was described in Section 4.3); the other consists of a single observer located at the origin who sends out consecutive light pulses which are reflected back to him from different points in space (Section 6).

8 Having established a method of assigning times to events, we can continue to make use of spacetime diagrams as we did in Block 1. As in the Newtonian case, each observing system has its own spacetime diagram. Towards the end of Section 6, it is shown how to construct world-lines on such diagrams.

Band 4 of AC2 comments on this Unit.

9 In Section 7, the question of simultaneity is revisited, this time being analysed with the help of spacetime diagrams.

Appendix A
The Michelson–Morley experiment

The famous optical interference experiment carried out by Albert Michelson in 1881 and subsequently with improved accuracy by Michelson and Morley in 1887, may be said to have sounded the death knell of the ether theory of light propagation (see the beginning of Section 3.2) which had dominated physical thought throughout the latter half of the nineteenth century. This was in spite of the fact that the experiment had been designed to *support* this theory, by detecting the motion of the Earth through the ether (the so-called 'ether drift').

If light propagates with the same speed c in all directions when measured in that special frame of reference which is stationary with respect to the ether, then on the basis of the Galilean velocity transformation, which until Einstein's time was quite unchallenged, we must expect the speed as measured in any frame which moves with respect to the ether to depend on the direction of propagation. To be specific, suppose that at some instant the Earth has velocity $\mathbf{v}$ *with respect to the ether*.[*] Then light propagating parallel to $\mathbf{v}$ should, according to Newtonian ideas, do so with speed $(c - v)$ as measured in a terrestrial laboratory. Similarly, the speed of light moving antiparallel to $\mathbf{v}$ should be measured in the moving laboratory as $(c + v)$ and that of light moving perpendicular to $\mathbf{v}$ in the laboratory should be $\sqrt{c^2 - v^2}$.

Objectives 2, 3, 8   **SAQ 28**   Derive this last result under the assumption of the previous paragraph that the Earth-bound laboratory moves at speed $|\mathbf{v}|$ relative to a coordinate system fixed to the ether. In this 'ether system', the speed of light is c.

Objective 8   **SAQ 29**   The best estimate for the speed of light in 1879 was the value $c = 299\,910 \pm 50\,\mathrm{km\,s^{-1}}$ obtained by Michelson himself. Would it have seemed at all feasible at that time to have attempted to detect the motion of the Earth through the ether by simply making direct, separate measurements of the speed of light along two opposing directions and seeing whether or not they differed?

The speed of the Earth about the Sun is approximately $30\,\mathrm{km\,s^{-1}}$.

The basic problem to which Michelson addressed himself was to discover whether two beams of light travelling different paths in different directions end up (on reflection back to their starting point) with one beam slightly lagging behind the other. Michelson devised an extremely clever method for detecting any such lag by using a phenomenon called interference. Interference arises when two wave trains are superposed. The resultant or net disturbance depends critically on whether the peaks and troughs of one wave train are superposed respectively on the peaks and troughs of the other wave train, thus giving a big net disturbance, or whether the trains

[*]You might argue that the Earth could be at rest with respect to this particular inertial frame. Quite possibly, but if so it will certainly not be at rest six months later, say, when the direction of its orbital motion has reversed. A modification of the Michelson–Morley experiment was carried out by Kennedy and Thorndike in 1932 in which they indeed took readings at intervals separated by *months*. See Appendix B4.

are superposed out of step so that, for example, the peak of one train coincides with a trough of the other and they tend to cancel one another out. When two light waves are superposed they give a bright resultant light if they are in step ('constructive interference') and darkness if they are out of step ('destructive interference'). Typically in an interference experiment, there are regions of both constructive and destructive interference. The pattern of alternating light and dark regions so produced is called an interference pattern and the light and dark regions are called interference fringes.

It is not necessary for you to understand the full details of Michelson's method; you only need to know that he used the technique of interference to detect any lag between one light beam and another. The reason he used this method is that it is exceedingly sensitive. A lag of half a wavelength corresponds to the difference between a bright interference fringe and a dark one. For visible light, half a wavelength corresponds to a distance of only about 3×10^{-7} m, and in practice one can detect smaller distances than that because one can differentiate between shades of brightness.

At the time the experiment was performed, one might have been forgiven for questioning whether the motion of the Earth through the ether could ever be detected. Indeed, Maxwell himself had regretfully concluded that it was beyond experimental capabilities. Michelson, however, responded to this challenge and the experiment he ultimately devised with Morley proved to be sufficiently sensitive to test the ether hypothesis. The experiment was made possible by Michelson's invention of an ingenious and versatile interferometer that now bears his name. The Michelson interferometer is shown schematically in Figure 16 which is based on the first figure of Michelson's original (1881) paper describing the instrument (*American Journal of Science*, **22**, 20).

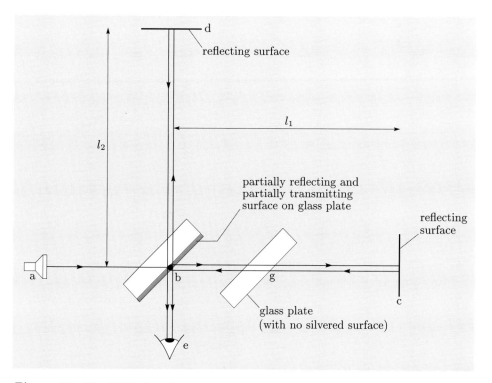

Figure 16 The Michelson interferometer

The first purpose of the instrument is to divide light from an extended source into two beams by partial reflection. These beams are sent in different directions (normally at right angles to each other), reflected against mirrors and brought back together again to form interference fringes. To quote Michelson:

> The conditions for producing interference of two pencils of light which had traversed paths at right angles to each other were realized in the following simple manner.
>
> Light from a lamp [a] passed through the plane parallel glass plate b, part going to the mirror c, and part being reflected to the mirror d. The mirrors c and d were of plane glass, and silvered on the front surface. From these the light was reflected to b, where one was reflected and the other refracted, the two coinciding along be.
>
> The distance bc being made equal to bd, and a plate of glass g being interposed in the path of the ray bc, to compensate for the thickness of the glass b, which is traversed by the ray bd, the two rays will have travelled over equal paths and are in condition to interfere.

[A. A. Michelson (1881) *Am. J. Sci.*, **22**, 20]

Plates b and g are set at 45° to the incident beam, so the reflected and transmitted beams travel mutually perpendicular paths. In subsequent experiments, including the Michelson–Morley experiment and most present-day applications, the back surface of plate b is only lightly silvered, so that the two beams arriving at e have equal intensity. Note that only those beams essential to the functioning of the apparatus are indicated in the simplified Figure 16. The apparatus of Michelson and Morley (1887) differed from that of Figure 16 in that *several* reflections from mirrors like c and d were incorporated in order to lengthen the effective paths bd and bc and thus improve the experiment's accuracy. A contemporary photograph of the apparatus is shown in Figure 17.

The idea behind the experiment becomes clear if we calculate the times taken for the light to travel from b to the mirrors c and d and back to b again on the ether hypothesis.

Figure 17 The Michelson–Morley experiment (from a photograph found between the pages of Michelson's laboratory notebook for the years 1891–95)

Suppose that the apparatus is oriented so that the direction bc is directly into the 'ether wind', of speed v. We readily see from Figure 16 and SAQ 28, *using the Galilean velocity transformation*, that the time t_{bcb} and t_{bdb} to travel bcb and bdb are respectively:

$$\left.\begin{aligned} t_{bcb} &= \frac{l_1}{c-v} + \frac{l_1}{c+v} = \frac{2l_1/c}{1-v^2/c^2} \\ t_{bdb} &= \frac{2l_2}{\sqrt{c^2-v^2}} = \frac{2l_2/c}{\sqrt{1-v^2/c^2}} \end{aligned}\right\}. \tag{A1}$$

The *difference* in travel time for the two beams arriving at eye e is thus:

$$\Delta t = t_{bcb} - t_{bdb} = \frac{2l_1/c}{1-v^2/c^2} - \frac{2l_2/c}{\sqrt{1-v^2/c^2}}. \tag{A2}$$

If one turns the apparatus through an angle of 90° so that bd now points along the direction of motion, into the 'ether wind', we find the time difference is now

$$\Delta t' = t'_{bcb} - t'_{bdb} = \frac{2l_1/c}{\sqrt{1-v^2/c^2}} - \frac{2l_2/c}{1-v^2/c^2} \tag{A3}$$

where we interchange denominators in Equation A2.

The time difference Δt for the two beams arriving at e will give rise to a set of interference fringes at e. If the interval Δt is changed for some reason, by adjusting the relative lengths l_1 and/or l_2 for example, then the pattern of interference fringes is observed to *shift* since, of course, the relative phase of the two beams arriving at e will have been altered.

Some typical interference patterns observed when viewed along eb are shown in Figure 18. The curvature of the fringes can be changed by minor adjustments of the orientation of mirrors c or d (one or both of these mirrors are fitted with adjusting screws). Michelson and Morley actually used fringes produced by a source of white light and these fringes are only visible when l_1 and l_2 are equal to within a few wavelengths.

As we have said, changing the relative phase (or 'step') of the two beams at e causes a shift in the fringe pattern. Such a change could be achieved, for example, by moving mirror c towards b, so that the time t_{bcb} is reduced by an amount δt, say. Since the wave travels at speed c (we neglect the small correction implied by the Galilean transformation of $\pm v$), the same difference δt corresponds to a shift of $\pm c\,\delta t/\lambda$ wavelengths in the path travelled by this beam (where λ is the wavelength of light).

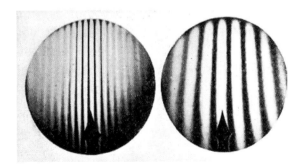

Figure 18 Typical fringe pattern observed in the Michelson interferometer

Another way of changing the relative phases of the two beams, according to Equations A2 and A3, is to rotate the apparatus. For on rotating through 90°, the time difference between paths bcb and bdb *changes* by

$$\delta t = \Delta t' - \Delta t$$

and we expect a corresponding fringe shift between the two beams of

$$\frac{c\delta t}{\lambda} = \frac{c}{\lambda}(\Delta t' - \Delta t) \quad \text{wavelengths.}$$

Since we may confidently expect the speed v of the Earth through the supposed ether to be very much less than the speed of light c in the ether, we may obtain a sufficiently good approximation to Δt and $\Delta t'$ by expanding them in a series in increasing powers of $(v/c)^2$. The details are unimportant here. The result is

$$\text{fringe shift } = \left| \frac{c\delta t}{\lambda} \right| = \frac{(l_1 + l_2)}{\lambda} \frac{v^2}{c^2}$$

$$+ \text{ negligible terms of order } \frac{v^4}{c^4} \text{ and higher. (A4)}$$

Objective 8

SAQ 30 Estimate from Equation A4 the fringe shift expected by Michelson for his first experiment (at Potsdam, in April 1881) with $l_1 = l_2 = 1.2\,\text{m}$, and by Michelson and Morley (Cleveland, July 1887) with $l_1 = l_2$ increased by successive reflections to an effective $11\,\text{m}$.

Take $v = 30\,\text{km s}^{-1}$, $\lambda = 570 \times 10^{-9}\,\text{m}$ for visible light, and $c = 3 \times 10^5\,\text{km s}^{-1}$.

Michelson and Morley were able to achieve considerable mechanical and thermal stability for their apparatus and were confident of its ability to detect fringe shifts as small as 0.01 of a fringe or step when it was rotated through 90°. It should therefore have been easy to observe the 0.4 fringe shift expected on the basis of Equation A4. *No such shift was observed.* On completion of these experiments, Michelson wrote to Lord Rayleigh,

R. S. Shankland (1964) *American Journal of Physics,* **32**, 16. This article provides a very readable account of the Michelson–Morley experiment.

> The experiments on the relative motion of the earth and ether have been completed and the result decidedly negative. The expected deviation of the interference fringes from the zero should have been 0.40 of a fringe — the maximum displacement was 0.02 and the average much less than 0.01 and then not in the right place.

> As displacement is proportional to squares of the relative velocities it follows that if the ether does slip past the relative velocity is less than one sixth of the earth's velocity.

The modern-day interpretation of the null result of the Michelson–Morley experiment is to regard it as a convincing demonstration that the speed of light always has the same value independent of the direction of propagation (see also Appendix B3), for clearly there will be no time difference between beams travelling two interferometer paths of *equal lengths* if they do so with *equal speeds*.

Appendix B   Experimental evidence for the constancy of the speed of light

As discussed in Section 3.3, we believe that the speed in free space of a light signal (or, more generally, *any* electromagnetic signal) has the same value for all time, independent of its frequency or the nature and motion of its source, independent of the direction of propagation or of the time and place of measurement, and independent of the particular inertial frame with respect to which this speed is measured. The fundamental importance of the principle of the constancy of the speed of light (Postulate SR2) in determining our conception of time is emphasized in Section 4. With time entering into the expression of virtually all physical laws, this principle is certainly a cornerstone of modern physics.

Many years of measuring the speed of light at a wide variety of wavelengths culminated in 1986 with the elevation of the constancy of the speed of light to an even more central status. In that year, CODATA, the Committee on Data for Science and Technology of the International Council of Scientific Unions published a report giving a recommended and self-consistent set of fundamental physical constants (i.e. not only c, but $\hbar$, the mass of the electron, and many more.) A noteworthy feature was that c was given as $299\,792\,458\,\mathrm{m\,s^{-1}}$ *exactly*. One consequence is that the metre is now fixed in terms of the second. In this appendix, we explore some of the history of experiments which have led physicists to be so confident in the principle of the constancy of the speed of light. We have already seen evidence that the speed of light is independent of direction of propagation in Appendix A.

B1   The speed of light

The first successful determination was that of Römer who in 1676 found that $c = 214\,300\,\mathrm{km\,s^{-1}}$. This value, obtained by using observations of Jupiter's satellites that effectively timed the passage of light across a chord of the Earth's orbit, differs from later (more accurate) estimates by about 30%. The accuracy of this first determination contrasts sharply with the most modern techniques, which by 1982 (i.e. prior to CODATA's recommendation) had achieved a precision of some 0.004 parts per million. These measurements were based on independent experiments of the frequency (f) and wavelength (λ) of radiation from specially stabilized lasers, with c being obtained by multiplication ($c = f\lambda$). The principal uncertainty was in the wavelength measurement.

B2   Constancy with respect to time, place and frequency

The fact that the speed of light has remained apparently constant for a considerable period of time, and is independent of the place where the measurements are carried out, is established by a simple comparison of measurements of c made at various places over the years. A brief and partial summary is presented in Table 1. Note that the further back we go in time, the less accurate are the estimates, so it is difficult to assign very accurate limits to the time variation of c; this table also gives you an idea how the precision increased between the 1840's and 1960's.

TABLE 1   The speed of visible light as measured over the years

Year	Investigator	Speed of light/$km\,s^{-1}$
1849	Fizeau	313 000
1862	Foucault	298 000 ± 500
1879	Michelson	299 910 ± 50
1926	Michelson	299 796 ± 4
1955	Schöldström	299 792.4 ± 0.4
1967	Karolus	299 792.4 ± 0.2

The determinations of Table 1 were all made using *visible* light. What evidence is there that the speed of electromagnetic radiation is independent of frequency?

Table 2 presents some, now quite old, measurements covering a wide range of frequencies. We see that in the range from 2.4×10^{10} Hz (microwaves, wavelength 12 mm) to 5.4×10^{14} Hz (visible light, wavelength 560 nm) the measured speed is constant to about one part in 10^6. To an accuracy of 1 in 10^5, the range can be extended down to 1.7×10^8 Hz (radio waves, wavelength 1.8 m) and to an accuracy of 1% the range can be extended up to 4×10^{22} Hz (γ-radiation, wavelength 7.3×10^{-6} nm). To the data of Table 2, we could add the value $c = 299\,790 \pm 30$ $km\,s^{-1}$ obtained by Rosa and Dorsey in 1907, employing only *static* electric and magnetic fields and evaluating c from Equation 3 of this Unit. This would not only add a zero-frequency entry to Table 2 but also give added support to Maxwell's theory of electromagnetism which itself predicts c to be independent of frequency.

TABLE 2   The speed of electromagnetic waves of different frequencies

Investigator	Frequency/Hz	Speed/$km\,s^{-1}$
Luckey & Weil	4.1×10^{22}	297 000 ± 3000
Cleland & Jastram	1.2×10^{20}	298 000 ± 2000
Schöldström	5.4×10^{14}	299 792.4 ± 0.4
Rank, Bennett & Bennett	10^{10}–10^{14}	299 794 ± 2
Froome	2.4×10^{10}	299 793.0 ± 0.3
Florman	1.7×10^8	299 795 ± 3

Finally, some precise comparisons of the speeds of electromagnetic waves of different frequencies have been made by studying events that give rise to the simultaneous emission of waves of different frequencies. For example, experiments on short pulses of visible light and high-energy γ-radiation, generated in the 3 km linear accelerator at Stanford, USA, have shown their velocities to be equal to within 6 parts in 10^6. Because of the large distances involved, time measurements of astronomical events offer attractive possibilities for precision determination of the relative speeds of electromagnetic waves of different frequencies. Thus, observations of visible (5.5×10^{14} Hz) and radio (2.4×10^8 Hz) waves from stellar flares shown their speeds to be equal to within one part in 10^6, while observations on visible pulses from the Crab Nebula pulsar have shown the speed of light to be constant to about 5 parts in 10^{18} over the visible range of frequencies.

B3 Constancy with respect to direction of propagation: the Michelson–Morley experiment

This famous experiment has been described in detail in Appendix A. Here we merely point out that it not only virtually rules out the ether hypothesis, but also the anisotropy of space. If free space were anisotropic with respect to the propagation of electromagnetic disturbances, then one would expect a fringe shift when the apparatus is rotated. In special relativity, the assumption of isotropy of space (Assumption A4 of Block 1) is carried over from the Newtonian world-view.

B4 Constancy with respect to frame of reference: the Kennedy–Thorndike experiment

As described in Appendix A, Michelson and Morley used an interference apparatus with arms of *equal* length. They rotated their apparatus over a period of minutes, which was a short time compared with the time for their Earth-based laboratory significantly to alter its velocity about the Sun. Thus, a single measurement demonstrated that a particular inertial frame is isotropic with regards to the propagation of light. By carrying out such measurements at various times over a year they showed that each of several distinct inertial frames is similarly isotropic. It was possible, however, that the actual speed of light might vary from one inertial frame to another. It turns out that because their apparatus had arms of equal length, they could not check this.

This possibility was however ruled out in 1932 by an adaption of the Michelson–Morley apparatus. The apparatus was based, like that of Michelson and Morley, on the very sensitive interference technique described in Appendix A, but it differed in two ways. First, the two arms were of different length (by about 160 mm). Second, they endowed the apparatus with such mechanical stability that measurements might be extended over periods of several *months*. The unequal arms meant that if the velocity of light were to change, there would be a consequent change in the interference pattern; this was not seen.

No fringe shifts were observed over a period of months. This implied that the time taken for light to travel the extra distance in one arm is the same in reference frames having different relative speeds; in other words, the speed of light is the same for all inertial frames.

B5 Constancy with respect to motion of the source

An accelerator experiment

The Michelson–Morley experiment was one in which the source of light was stationary relative to the observer. That the speed of light is independent of the motion of its source has been demonstrated by direct measurements of the speed of electromagnetic radiations emitted by rapidly moving unstable particles produced in very high-energy accelerators. For example, high-energy radiation emitted in the decay of neutral π mesons, themselves having a speed very close to that of light, were measured in the laboratory to have a speed of $(2.9979 \pm 0.0004) \times 10^8 \, \mathrm{m\,s^{-1}}$, in excellent agreement with the accepted value of c.

An experiment with starlight

This experiment is considered at some length on the videotape. Here we give you a fairly detailed account of an experiment performed on the electromagnetic radiation received from a distant binary star system (two stars revolving about their common centre of mass).

It was pointed out by de Sitter in 1913 that binary stars provide a fast moving source of light, and that their appearance depends on whether the speed of their light towards us is affected by their motion. To take a simple case first, consider a pair of stars in mutual circular orbits whose plane passes through the centre of the Earth. Choose one star for observation. Call its speed v. If light behaves like a bullet (a 'ballistic' theory), then light from the star when at one extremity of the orbit (point B in Figure 19), travels to us with speed $(c + v)$ and when the star is on the other side, at point D, the light travels to us with speed $(c - v)$.

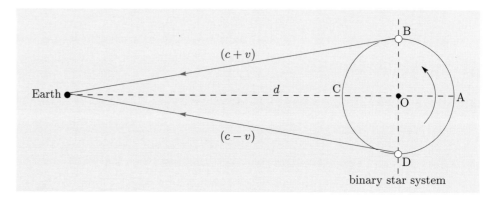

Figure 19   A binary star system. Point O marks the common centre of mass of the two stars about which both revolve. B and D are two points on the orbit of the star whose light is observed. The second star does not appear in this figure.

The time taken for the light to reach the Earth is $d/(c + v)$ in the first case and $d/(c - v)$ in the second, where d is the distance of the pair of stars from the Earth. One can imagine a case where the difference between these two times is just half the period of their mutual rotation. In this case, the star would be *seen* on both sides of its companion simultaneously — the system would look like three stars. More generally, in cases where the time difference of the light arriving simultaneously at the Earth from different parts of the star's orbit is not exactly half a period, one would still get multiple or smeared-out images. Assuming that the velocity of light from a source moving towards the observer with speed v is given by $(c + kv)$, where k is a constant that would have the value unity in a simple ballistic model, it was concluded that the non-existence of multiple images requires k to be less than 10^{-5}.

Actually, one has to be a little careful with this argument. Between the binary star and us is interstellar gas. We know that visible light is affected when passed through a gas — it is refracted. In the early 1960s, it was pointed out that with visible light one might get misleading results. It could be argued from something called the 'extinction theorem' that even if the light were emitted from the star with a high or low speed due to the source moving, it would quickly alter its speed to the normal value as it passed through the stationary intervening gas. Whether this objection is valid is not entirely clear. However, we know that X-rays are largely unaffected by their passage through gas. So, to be on the safe side, the binary star experiment has been done with stars emitting X-rays, and the maximum value quoted above, $k = 10^{-5}$, was actually obtained from observations on such stars.

In point of fact, one can do even better than this if one is prepared to make a careful study of the apparent frequency of the X-rays being received. There are two reasons why the frequency of the X-rays as observed at the Earth could differ from that at emission. The first is the normal Doppler effect. This is due to the fact that light pulses are more bunched up when the source moves towards us and more spread out when the source recedes. So the usual Doppler effect would say that for any point on the half of the orbit ABC (see Figure 20), where the star is coming towards the Earth, the apparent frequency will be higher than normal, whereas for points on the half CDA the frequency will be lower. Indeed, one would expect a sinusoidal variation as in Figure 21. Note that this effect should occur regardless of whether the speed of light depends on the motion of the source or not.

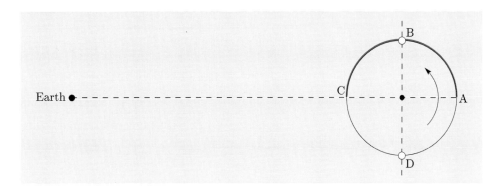

Figure 20   Four points A, B, C and D on the orbit of the observed star about the centre of mass of the binary system

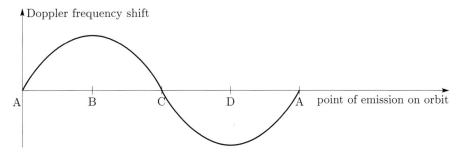

Figure 21   The Doppler shift for the same four orbital points as on the previous figure

If the speed of light depends on the motion of the source, then the frequency would be further modified. If we consider two successive pulses emitted towards the Earth, the second emitted with a somewhat lower speed than the first, then the distance between the pulses will increase as they journey through space, and the observer on Earth will measure a reduced frequency. This will apply over the half of the orbit BCD (see Figure 22). It is around this half that the speed of the source as directed towards the Earth is continually decreasing from a maximum at B, to zero at C, and to a maximum in the opposite direction at D.

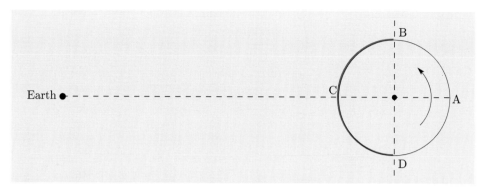

Figure 22   Four points A, B, C and D on the orbit of the observed star about the centre of mass of the binary system

Around the other half, DAB, the speed towards the Earth is increasing and the apparent frequency is also periodic, but as you can see from Figure 23, it is out of step with the usual Doppler variation. Combining the two effects gives a variation somewhat displaced from the usual Doppler curve. By measuring how much the resultant curve is displaced from that expected from the Doppler effect alone, one can estimate the contributions from variations in the speed light. In practice, no displacement has been found, and, referring back to our expression $c + kv$, the upper limit on k is now $k < 10^{-9}$. From this, we can indeed conclude that the speed of light does *not* depend on the speed of the source.

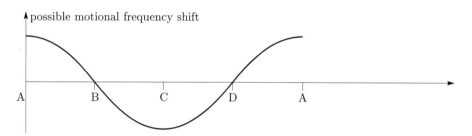

Figure 23   Possible motional frequency shift

Self-assessment questions — answers and comments

SAQ 1 Since the electron moves to the right in Figure 2, its direction of $\mathbf{v} \times \mathbf{B}$ points *away* from the wire as in Figure 24. But magnetic force is $q(\mathbf{v} \times \mathbf{B})$. Since, for an electron, q is a negative number, the direction of the magnetic force lies *towards* the wire; the electron is attracted towards the wire.

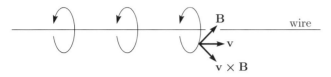

Figure 24   See SAQ 1

SAQ 2 We have conjectured only that the *form* of Equation 1 is the same for all inertial frames. Most generally, then, we ought to have written q', and not q, in Equation 2b. In assuming that the charge of the test electron is the same in both frames, i.e. that $q' = q$, we are inserting into the thought-problem of Section 2 the major assumption that charge is *invariant*; we are assuming that a charge has the same value for all inertial frames.

SAQ 3 The electric force is $q\mathbf{E}'$. We desire that the electron be attracted towards the wire. Thus the vector force $q\mathbf{E}'$ must point towards the wire. But, for an electron, q is a negative number. Thus $q\mathbf{E}'$ is antiparallel (i.e. parallel but in the opposite direction) to $\mathbf{E}'$. If the electron is to be attracted to the wire then $\mathbf{E}'$ must point away from the wire.

SAQ 4 According to our argument, the electron is attracted towards the wire in frame S'. Further, we have concluded that the source of this force is Coulombic. According to Coulomb's electrostatic law, opposite charges attract one another (Unit 4). Thus the wire must appear to have a net positive charge in frame S', if the electron is to be attracted to it.

SAQ 5 Since we have concluded that, in S', the wire has a net positive charge, Figure 3(b) ought to indicate more positive than negative charges per unit length. Figure 4(b) (on p. 8) has been drawn to show this. Notice that Figure 4(a) (or Figure 3(a) for that matter) has been drawn to indicate charge neutrality in frame S.

SAQ 6 Actually, we have argued only that there must be a *net* positive charge per unit length of the wire as observed in frame S'. A net positive charge means an excess of positive over negative charge. The ions appear to 'compress' when you speed up to move with the electrons; consequently, from the vantage point of motion with the electrons, the electrons would have to 'compress' when you speed up to move with the ions (i.e. change to the rest frame of the ions or wire). So the reverse of this latter operation, i.e. speeding up to move with the electrons, might be expected to space them out. In any case, we conclude that length is relative.

SAQ 7 We have made the following major assumptions:

1 A long wire carrying a steady current is electrically neutral in its rest frame. This is well attested experimentally.

2 The magnetic field outside a long wire carrying a steady current is as drawn in Figure 2 on p. 5. This has been verified experimentally.

3 The Lorentz force law, Equation 1, describes the force due to fields $\mathbf{E}$ and $\mathbf{B}$ acting on a test charge.

4 The Lorentz force law takes exactly the same form in any inertial frame.

5 Charge is invariant, i.e. if a test particle has charge q in inertial frame S, then it also has charge q in any other inertial frame S'.

SAQ 8 (a) The beam running parallel to $\mathbf{v}$ would have an *apparent* speed $(c - v)$ in the laboratory, while the beam running antiparallel to $\mathbf{v}$ would have an apparent speed of $(c + v)$ in the laboratory.

(b) For $v = 30 \, \mathrm{km\,s^{-1}}$ the net shift in apparent light speed would be $60 \, \mathrm{km\,s^{-1}}$.

SAQ 9 It appears that loss of simultaneity in frame S' is due to one end of the rod 'running away' from its pulse while the other end 'runs up' to meet its pulse. If the rod's length approached zero, this effect would have no time to develop in any frame; the pulses would strike the ends of a rod of infinitesimal length simultaneously *for all inertial frames*.

SAQ 10 Light signals, indeed electromagnetic disturbances of any type, don't depend on any medium for their propagation. The speed of electromagnetic waves in free space, as measured in any inertial frame, has the same value c, independent of their frequency, wavelength, amplitude, direction of propagation, and motion of the source. In fact, it will turn out that a most important aspect of electromagnetic waves is that they have the *highest* speed of propagation of any signal. If there were any other signal that moved faster than light, then this would be preferable. If there were a signal that moved infinitely fast, then time would be absolute in the old Newtonian sense. This non-physical hypothetical situation is briefly considered in the worked example at the end of Section 4.3.

SAQ 11 Measurements of the velocity of light involve the measurement of time intervals at a single position in space. Such measurements involve sending a beam of light out from a spatial point and then reflecting it back to the same point. No difficulties are experienced with the Newtonian concept of time if we are only concerned with measuring time intervals between events at the *same place* (we can imagine a clock located at that place if you like). Einstein's contribution is concerned essentially with the assigning of times to events *separated* in space.

SAQ 12 To satisfy the first requirement, within any given inertial frame it must be possible to assign a unique, well-defined time to any event. Clearly if, as discussed in the preceding paragraph, two observers at different points in the same frame cannot agree that all clocks are synchronized, then this condition is not met.

If the speed of light were infinite then the 'delay times' for all observers in the given inertial frame would be zero; all such observers would now agree that the clocks are synchronized. This is a *non-physical* situation.

SAQ 13 In *both* cases, the procedure is simply a numerical statement of the *same* fundamental hypothesis, namely that light travels at the same speed independent of its direction of propagation; this means, in particular, that equal distances backwards and forwards along a given line are travelled in equal times.

SAQ 14 Referring to Figure 8, we have $t_B - t_A = AB/c$ and $t'_A - t_B = BA/c$. Now consider propagation of light from B at time T_B to A at time T_A and back to B at time T'_B. We have $T_A - T_B = BA/c = AB/c$, and $T'_B - T_A = AB/c$. Thus, we have

$$T_A - T_B = T'_B - T_A.$$

This has the form of Equation 4a, but with A and B interchanged. Thus, if Equation 4a describes the synchronization of clock B with clock A then the above equation means that clock A is synchronous with clock B.

SAQ 15 Given that B and C have both been synchronized with A, we can assign the 'A times' to events at A, B and C. Let a beam of light leave A at time t_A, be reflected at B towards C at time t_B, be reflected from C towards A at time t_C and arrive back at A at time t'_A (Figure 25). The overall distance travelled by the beam is $(AB + BC + CA)$. Now clocks A and B have been synchronized, and therefore

$$t_B = t_A + \frac{AB}{c}.$$

Similarly, clocks A and C have been synchronized, so

$$t'_A = t_C + \frac{CA}{c}.$$

Furthermore the speed of light is the same on all three legs, so we may say

$$t'_A - t_A = \frac{AB + BC + CA}{c}.$$

Combining these three equations gives

$$t_C = t_B + \frac{BC}{c}$$

which means that clocks B and C have been synchronized.

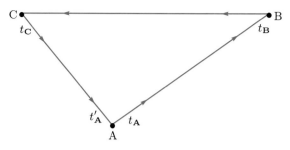

Figure 25   See SAQ 15

SAQ 16 Since the speed of light is the same in all directions, we need merely set all clocks at this distance from the origin at time $t_R = R/c$ if the pulse is initiated at time zero on the clock at the origin. We could synchronize all clocks in space by this method, because R can have any value.

SAQ 17 The world-line is shown in Figure 26.

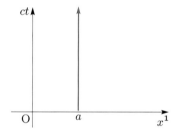

Figure 26   See SAQ 17

SAQ 18 The situation depicted in Figure 12 would be entirely inappropriate, because it indicates that some light signals existed for negative values of time. This is quite impossible since the pulse is created at time $t = 0$.

SAQ 19 Figure 13 apparently purports to show the world-line of a uniformly moving particle, but it is nonsensical since the slope is less than 1, and therefore indicates that the particle is moving faster than c.

SAQ 20 The world-line of the light pulse reflected by a mirror at $x^1 = a$ is shown in Figure 27.

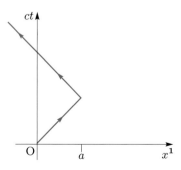

Figure 27   See SAQ 20

SAQ 21 The world-line of the pulse sent out at time t_{out} is shown in Figure 28.

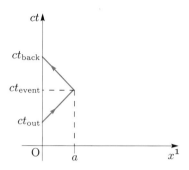

Figure 28   See SAQ 21

SAQ 22 Since the light pulse travels out and back at speed c, the observer can calculate the location of the event from the simple equation

$$(t_{\text{back}} - t_{\text{out}}) = \frac{2a}{c}.$$

In other words,

$$a = \frac{c(t_{\text{back}} - t_{\text{out}})}{2}.$$

SAQ 23 Speaking operationally, we may imagine seeding the relevant regions of space with data takers, each with an identical clock and each fixed to our chosen inertial frame. The clocks may by synchronized, for example by the procedure of SAQ 15. Each such data taker can then record the time, on his synchronized clock, of any events occurring in his immediate vicinity. The times and places of all events can later be collected.

An alternative and entirely equivalent method is to have a single observer at the origin, say, who can reconstruct the times and places of events distant from him by arranging things so that he can effectively 'trigger off' these events by means of coded light pulses.

SAQ 24 Strictly, 'observer' refers to any observing system, associated with any inertial frame, able to assign times and positions to events. Two ways of doing this were considered in the previous SAQ.

A spacetime diagram, for events occurring along the x^1-axis at least, is a two-dimensional plot of the times and locations of any events of interest. These spacetime diagrams look like those of the old Newtonian world-view, but the times and locations they record are now subject to the new special theory.

SAQ 25 The coordinates of event $\mathscr{E}_{\text{B}}$ are

$$(ct_{\text{B}}, x_{\text{B}}) = \left(\frac{c(t_2 + t_4)}{2}, \frac{c(t_4 - t_2)}{2} \right).$$

SAQ 26 According to O′, the rod moves uniformly to the right and the left-hand pulse reaches its end of the rod before the right-hand pulse reaches its end, as shown in Figure 29.

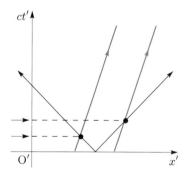

Figure 29   See SAQ 26

SAQ 27 Loss of simultaneity is due to the fact that the speed of light is finite and is the *same* for all inertial observers.

SAQ 28 In the laboratory system, the 'ether wind' has velocity $\mathbf{v}$. If the light beam is directed so as to move perpendicular to $\mathbf{v}$ then we have the vector diagram shown in Figure 30. In this Figure, $\mathbf{c}$ represents the velocity of the beam of light. By Pythagoras's theorem, we have

$$u^2 = c^2 - v^2 \quad \text{or} \quad u = \sqrt{c^2 - v^2}.$$

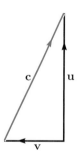

Figure 30   See SAQ 28

SAQ 29 Michelson's value for c was uncertain to within $\pm 50\,\text{km s}^{-1}$, so that one could not hope at that time to detect differences of the order of $30\,\text{km s}^{-1}$ between two measurements of this type.

SAQ 30 From Equation A4, for Michelson's (1881) Potsdam version, we have

$$\left| \frac{c\delta t}{\lambda} \right| \approx \frac{2 \times 1.2}{570 \times 10^{-9}} \left(\frac{30}{3 \times 10^5} \right)^2$$
$$\approx 0.042 \text{ of a fringe.}$$

The subsequent experiment of Michelson and Morley increased the length by a factor of $11/1.2$. Since all other parameters were the same, we now have

$$\left| \frac{c\delta t}{\lambda} \right| \approx \frac{11}{1.2} \times 0.042$$
$$\approx 0.38 \text{ of a fringe.}$$

The sensitivity was thus enhanced by about a factor of 10. Actually, many other improvements were made, the main one being greatly improved mechanical stability and isolation from outside disturbances. In 1887, the smallest shift that it was possible to observe, when accounting for all sources of error, was 0.01, so the 1887 finding was clearly a convincing null result. Subsequent improvements have strengthened this conclusion.

Acknowledgements

Grateful acknowledgement is made to the following sources for material used in this Unit:

Figure 5 from Clark University Archives; *Figure 6* from U.S. Library of Congress; *Figure 17* from Michelson Collection, Nimitz Library, U.S. Naval Academy, courtesy of Hale Observatories; *Figure 18* from *Reviews of Modern Physics*, **5**, No. 3, July 1933.

Unit 6 Some consequences of special relativity

Prepared by the Course Team

Contents

Aims

In this Unit we intend to:

1 Follow a line of reasoning leading to the famous Lorentz transformation formulae which lie at the heart of special relativity.

2 Give several detailed applications of the Lorentz transformation.

3 Emphasize the need for careful use of terms in special relativity by considering two effects that might at first glance be interpreted as paradoxical.

Objectives

When you have finished studying this Unit, you should be able to:

1 Write down the Lorentz transformation formulae, Equations 23a to 23e, explaining what they represent and showing that they reduce to the Galilean transformation in the 'non-relativistic limit' (the case of very low velocity).

2 Given the appropriate formula, interpret the relativistic Doppler shift.

3 Derive the length contraction formula, Equation 24, and explain what it means.

4 Derive the time dilation formula, Equation 26, and interpret its meaning.

5 Given the velocity addition formulae, Equations 29, 31 and 32, explain their physical meaning.

6 Describe the outcome (Section 3.4) of the thought-experiment of Section 2 of Unit 5, though not necessarily by means of formulae of your own derivation. (You may be given certain equations to interpret or manipulate.)

7 Describe the difference between the relativistic Doppler effect and the time dilation effect, and use them to exemplify the care needed in the use of the word 'observation'.

8 Describe the twin paradox and give an inkling as to why it is no paradox at all (the story is to be continued in Unit 7).

1   Introduction to Unit 6

Band 5 of AC2 introduces this Unit.

In Unit 5 we showed you that the old Newtonian world-view is inconsistent with certain physical observations. We then began a fairly systematic examination of Einstein's new concept of spacetime. Now in Unit 6 we pursue this further, in three stages.

In the first stage we derive the famous Lorentz coordinate transformation which replaces the old Galilean transformation of Unit 1. In the appropriate limit of low velocities, the Lorentz transformation is well approximated by the Galilean transformation.

In the second, somewhat longer, stage we present certain fundamental applications of the Lorentz transformation, leading to expressions for 'length contraction' and 'time dilation'.

In the third stage we step back a little and consider certain seemingly paradoxical predictions of Einstein's theory. We hope that this stage will give you a chance to consolidate your understanding of time and space in the new picture of the world.

4

2 The Lorentz transformation

2.1 Different inertial frames

An essential feature of special relativity is the way it connects the coordinates of an event $\mathscr{E}$ as recorded by two different inertial observers. Suppose that an observer O attributes coordinates (t, x^1, x^2, x^3) to some isolated event, i.e. occurrence in spacetime (you might think of something like a firecracker exploding at time t and position x^1, x^2, x^3). We want to know the coordinates (t', x'^1, x'^2, x'^3) of the *same* event as recorded by some other observer, O', whose coordinate system is moving uniformly at velocity **V** relative to O as in Figure 1. For simplicity we consider the x'^1-axis of O' to be parallel to the x^1-axis of O. (Notice that the motion of one frame with respect to the other must be uniform since we we have just said that we are looking at an event as recorded by 'two inertial observers'; if one frame accelerated with respect to the other, at least one observer would no longer be inertial.) We shall frequently meet pairs of frames with their corresponding axes parallel as in Figure 1; such pairs of frames are sometimes referred to as being in *standard configuration*.

We hope our notation does not cause confusion. As you know, we label the three coordinate axes of O by x^1, x^2 and x^3. Many people like to use the labelling x, y and z for these axes, but we have strong reasons in this Course not to do so — in fact in Unit 7 we underline the equal standing of time and space by labelling the 'time coordinate', ct, by x^0. The price we pay for this notation is that we must label the coordinates of observer O' by x'^1, etc. One pronounces O' as 'O-prime', x^1 as 'x-one', and x'^1 as 'x-prime-one', and so forth. Actually we shall be able, for much of the Block, to follow the conventions in Section 6 of Unit 5, thereby considerably simplifying the notation.

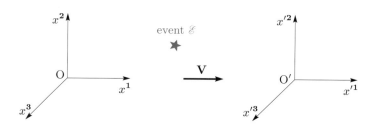

Figure 1   Two inertial observers record the same event $\mathscr{E}$ in spacetime. O' moves at velocity **V** relative to O and O moves at velocity $-\mathbf{V}$ relative to O'.

In the remainder of Section 2 we seek by stages (and find!) the famous Lorentz transformation connecting the coordinates (t, x^1, x^2, x^3) and (t', x'^1, x'^2, x'^3) of any event $\mathscr{E}$ for two such inertial observers whose relative motion is as depicted in Figure 1. As it happens, however, we may simplify things further by supposing for the main part of the analysis that event $\mathscr{E}$ is located *on* the collinear x^1- and x'^1-axes (see Figure 2). In this case we may suppose that O can specify $\mathscr{E}$ by assigning *one* space coordinate which, by the convention of Unit 5, we may call x, and the usual time coordinate, t. Similarly observer O' will label $\mathscr{E}$ by spacetime coordinates (t', x').

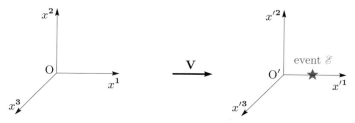

Figure 2   An event $\mathscr{E}$ on the common x^1- and x'^1-axes of O and O'.

Let us assume, for simplicity, that the observers have arranged things so that a special event whose coordinates according to O are $(t, x) = (0, 0)$ is assigned coordinates $(t', x') = (0, 0)$ by O'. (Here and henceforth we are using 'observer' in the sense of *observing system*, as discussed in Section 6 of Unit 5.) This is easily visualized, for we may think of O as located at his origin $(x = 0)$ with his clock, and O' at his origin $(x' = 0)$ with his identical clock, such that as O' (moving to the right in Figure 2) passes O they start their clocks set at times $t' = 0$ and $t = 0$ respectively. This is a simultaneous act, since at that instant both O and O' are located at the same point in space. You have seen in Section 7 of Unit 5 that simultaneity is a relative concept only for events occurring at different points in space. Finally (to repeat), although we have drawn x^2- and x^3-axes, and x'^2- and x'^3-axes in Figures 1 and 2, we consider the events to occur only on the x^1- and x'^1-axes, which by convention we shall call simply the x- and x'-axes.

Let us suppose we accept for the moment that $|\mathbf{V}|$ (the speed of O' relative to O) must be less than the speed of light, c. Then the spacetime trajectory (world-line) of the origin of O' will be for O as in Figure 3. Notice that the world-line must pass through the origin $(x = 0, t = 0)$ of O, since we have assumed that the respective origins of O and O' coincide at $t = 0$ (and $t' = 0$).

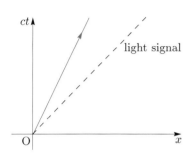

Figure 3 The world-line of the origin of O' as observed by O is shown in blue.

Objective 1

SAQ 1 You have encountered the pre-Einstein or Galilean transformation connecting the coordinates (t, x) and (t', x') of an event $\mathscr{E}$. This was discussed in Section 5.3 of Unit 1. By way of review, write this transformation down.

Unit 5 revision

SAQ 2 By considering SR2 (as in, e.g. Section 3.3 of Unit 5) give one reason why the Galilean transformation must fail to represent the relation between events as seen in O and O'.

With O' standing at his origin, $x' = 0$, but moving to the right relative to O, as in Figure 2, consider the following simple thought-experiment: observer O sends two pulses of light out towards O', separated by the time interval T on *his* clock at $x = 0$. The two pulses will catch up with O' and *he* will receive them separated by some time interval on *his* clock at $x' = 0$.

We make very nearly the *simplest assumption* that the time interval between the two pulses as received by O' (and recorded on his clock) is $k(V)\, T$, where we allow for the possibility that the factor k may be a function of V where V is directly related to the velocity $\mathbf{V}$ of O' relative to O. For velocities understood to be along one axis only (here the x-axis) we write V rather than $\mathbf{V}$. V is not the x-component of velocity, but a quantity whose magnitude is equal to the speed of O relative to O' and taking a positive value if O and O' are receding from each other, and a negative value if they are approaching one another.

Next we consider a similar experiment, but this time with observer O' sending out two pulses towards O separated by a time interval T on *his* clock. O' sees O to depart along the negative x'-axis.

Objective 1

SAQ 3   Draw the world-line of O on the axes (ct', x') of O'. (Remember that the origins of O and O' coincide at time $t' = 0$.)

Now comes a critical point of the argument: we draw on the first postulate of Unit 5, SR1, that all inertial frames are equivalent for expressing the laws of physics. Since O' sees O to recede at speed V (this is the speed, since here we chose our axes to be such that V would be positive), if he (O') sends out two pulses of light separated on *his* clock by time interval T, then O will receive them at an interval $k(V)\, T$ just as in the previous experiment. This follows because postulate SR1 says that the two inertial frames are *equivalent*, and postulate SR2 says that the speed of light is identical for O and O' and doesn't depend on the direction of its propagation in space. Figure 4 should make this clear. Note that we have implicitly assumed that k does not distinguish between the positive and negative x-directions, a natural assumption since there is no natural distinction between different directions of relative motion. However, k *does* depend on whether O and O' are receding from each other or approaching one another. (As we shall see later, $k(-V) \neq k(V)$.)

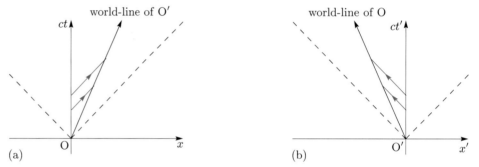

Figure 4   (a) For V positive, observer O sees O' depart to the right at speed V. Two light signals separated by T for O are received by O' with separation $k(V)\, T$.

(b) For the same relative motion, observer O' sees O depart to the left at speed V. This time, two light signals separated by T for O' are received by O with separation $k(V)\, T$.

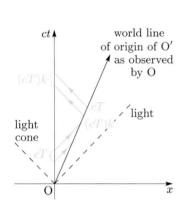

Figure 5   Two signals emitted from $x = 0$ at an interval $c(t_2 - t_1) = cT$ arrive at the moving point $x' = 0$ at an interval $c(t'_2 - t'_1) = (cT)k$. Note that the scale of lengths along the world-line $(x' = 0)$ is not the same as the scale of lengths along the ct-axis, $x = 0$. Two signals emitted from $x' = 0$ at an interval cT arrive at $x = 0$ at an interval $(cT)k$.

The two experiments are depicted together in Figure 5, as they appear to observer O. Note that we have chosen to use the axes (ct, x), but axes (ct', x') would have been equivalent. We think you will begin to see, from Figure 5, that a spacetime diagram can contain quite a lot of information. In particular, we emphasize the following features of Figures 4 and 5:

1 We choose to label the respective time axes by ct and ct' and not just t and t'. This is quite allowable, since c is a universal constant for all inertial observers. Of course, ct has the same dimensions as x.

2 A light pulse, with this choice of time axis, has a world-line at 45° to the x-axis (see answer to SAQ 19 in Unit 5). We often choose, henceforth, to indicate the spacetime paths of light pulses on our spacetime diagrams. These paths moving up on the left and right are universally called *light cones* for a reason which will be clearer when the y-, i.e. x^2-direction is also considered.

3 If the interval in time between two flashes is T, then on the ct-axis it is represented by cT.

In order to derive an expression for $k(V)$, we consider an event $\mathscr{E}$, occurring at the *origin* of O$'$ (that is, at $x' = 0$). Thus to O$'$, $\mathscr{E}$ has coordinates $(ct', 0)$. What are the coordinates of $\mathscr{E}$ according to O? As we have discussed in Section 6 of Unit 5, we may consider that O at his origin sends out a pulse at time t_1 on his clock, which triggers event $\mathscr{E}$ and then is reflected back to him at time t_3 on his clock.

This is depicted in Figure 6.

Objective 1   SAQ 4   Show that the coordinates (ct, x) of event $\mathscr{E}$ as seen by observer O are

$$(ct, x) = \left(\frac{c(t_1 + t_3)}{2}, \frac{c(t_3 - t_1)}{2} \right).$$

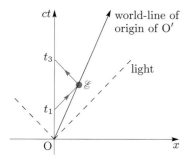

Figure 6   A signal emitted from O to O$'$ and reflected back to O.

Notice that as well as using ct- and x-axes on diagrams, we also refer to coordinates (ct, x) when it is appropriate. From the answer to SAQ 4, it follows that if we want to express the coordinates (ct, x) of $\mathscr{E}$ as seen by O in terms of the coordinates $(ct', 0)$ as seen by O$'$, we must express t_3 and t_1 in terms of t'. This is possible from our previous thought-experiment with time intervals (see Figures 4 and 5); since O and O$'$ pass one another at time $t = 0 = t'$, the first instant of the time interval is just $t = t' = 0$ and so we can write the time interval $t_1 - 0 = t_1$ in terms of the time interval $t' - 0 = t'$ simply as $t' = k(V)t_1$. Similarly, the time intervals $t_3 - 0$ and t' are connected by the equation $t_3 = k(V)t'$. And from these two equations we have:

$$t_3 = k(V)t' = k(V) \times k(V)t_1 = k^2(V)t_1.$$

But (from SAQ 4) the time (on the clock used by O at his origin) of $\mathscr{E}$ is $\frac{1}{2}(t_1 + t_3)$; so we have:

$$t = \tfrac{1}{2}(t_1 + t_3) = \tfrac{1}{2}(t_1 + k^2(V)t_1)$$
$$t = \tfrac{1}{2}(1 + k^2(V))t_1. \tag{1}$$

Similarly we have, from SAQ 4,

$$x = \frac{c}{2}(t_3 - t_1) = \frac{c}{2}(k^2(V)t_1 - t_1)$$
$$x = \frac{c}{2}(k^2(V) - 1)t_1. \tag{2}$$

Finally we can eliminate t_1 in Equations 1 and 2 by recalling that, since event $\mathscr{E}$ occurs at the origin ($x' = 0$) of O$'$, and since O$'$ moves to the right (as seen by O) with speed V, we have

$$x = Vt. \tag{3}$$

Notice that we do not have to use the longer expression $x = Vt + a$, where a is some constant, since, for simplicity, we have been considering the case such that the trajectory of O$'$ must pass through the point $(ct, x) = (0, 0)$ as in Figures 4 and 5. Since $x = 0$ when $t = 0$, by substituting these values for x and t in the equation $x = Vt + a$ it can be seen that $a = 0$. And since $\mathscr{E}$ is taken to lie at the origin of O$'$, it must therefore lie on the trajectory of O$'$ as seen by O.

It is now an exercise in algebra to show from Equations 1, 2, and 3 that

$$\frac{V}{c} = \frac{k^2(V) - 1}{k^2(V) + 1} \quad \text{and} \quad k(V) = \sqrt{\frac{1 + \dfrac{V}{c}}{1 - \dfrac{V}{c}}}. \tag{4}$$

(You may want to attempt this yourself before reading our analysis below.)

Briefly, dividing Equation 2 by Equation 1 gives

$$\frac{x}{t} = \frac{\frac{1}{2}c(k^2(V) - 1)t_1}{\frac{1}{2}(k^2(V) + 1)t_1} = c\left(\frac{k^2(V) - 1}{k^2(V) + 1}\right).$$

But, from Equation 3, $x/t = V$. Thus

$$V = c\left(\frac{k^2(V) - 1}{k^2(V) + 1}\right)$$

which is equivalent to the first form given in Equation 4. To derive the second form in Equation 4, multiply both sides of the first form by $(k^2(V) + 1)$ to get

$$\frac{V}{c}(k^2(V) + 1) = k^2(V) - 1.$$

Then collect terms in $k^2(V)$ to find

$$k^2(V)\left(1 - \frac{V}{c}\right) = 1 + \frac{V}{c}.$$

The second form in Equation 4 follows by dividing by $(1 - V/c)$ and taking the square root. In this step, retaining only the *positive* square root makes physical sense.

Let us try to summarize our results so far. (We suggest that you might try to do so independently as an exercise before reading further.)

We have considered two inertial frames, as in Figure 2, receding from one another uniformly with speed V. If *either* observer sends out, from a fixed point in his frame, two light flashes separated by time interval T on his clock, then we have reasonably assumed that the other observer will receive the pulses separated by an interval $k(V)\, T$ on his clock (which in general we would expect to be different from T) at a point fixed in his frame. And the postulates of relativity have led to Equation 4 for $k(V)$. Throughout our analysis we have thought of V as positive. Since V is the velocity of O′ *as observed by* O, this means that we have assumed that O′ recedes from O and vice versa. However, we have carefully not assumed that V *must* be positive. Thus, Equation 4 holds for V being negative, when O and O′ approach one another. We *have* assumed, however, that the speed $|V|$ is less than c, for otherwise no light signal sent by one observer could be observed by the other for the cases depicted in Figures 4(a) and 4(b).

By now, you can see how the relativity of time is coming out of our assumptions.

Objectives 2 and 7   **SAQ 5**   Show that if $V = 0$ then $k(V) = 1$, that if $0 \leqslant V/c < 1$ then $k(V) \geqslant 1$, and that if $-1 < V/c < 0$ then $k(V) < 1$.

Objectives 2 and 7   SAQ 6   Sketch a rough graph of k as a function of V/c for all values of V/c between -1 and $+1$. Is $k > 1$ when O and O' are receding or approaching?

2.2   The relativistic Doppler shift

Before proceeding to derive the full Lorentz equations relating to the coordinates (t, x) and (t', x') of an *arbitrary* event $\mathscr{E}$ which is not necessarily located at the origin of O or of O', let us consider a physical application of Equation 4 for $k(V)$. This application describes an effect known as the *relativistic Doppler shift* (it is also sometimes known as the *Doppler shift for light*, or just *Doppler shift*).

The *acoustic* Doppler effect is very familiar; you must have noticed that the sound of an ambulance siren appears to have a higher frequency as the ambulance approaches and then a lower one as it recedes. This effect can be explained accurately in terms of simple Newtonian physics by considering that sound waves move through the medium, air, at about $300 \, \mathrm{m \, s^{-1}}$.

The situation is very different for the relativistic Doppler shift. In the first place, the acoustic Doppler shift occurs for sound propagation in a *medium*, but electromagnetic signals require *NO* medium for their propagation. Indeed, the acoustic effect is described by two *different* equations according to whether the sound source or receiver moves with respect to the medium. According to postulate SR1, however, we cannot draw such a distinction between any two inertial observers.

Let us visualize the simplest case (Figure 2) where two observers are approaching or receding along the line between them. If O, say, transmits a steady light of frequency f towards O' along their mutual x-axes, what is the frequency f' of the received light according to O'? Tentatively we might suppose that if O and O' are receding from one another, then O' will say that the frequency of the light is lower than f. This means the light will be shifted towards the red end of the spectrum. Similarly, if O and O' are approaching one another, the frequency will be raised and the light will be shifted towards the blue end of the spectrum. Of course, as discussed in Unit 5, instead of visible light we might have radio waves, for example. All such electromagnetic disturbances are governed by Maxwell's wave equation and propagate with speed c in empty space.

The analysis of Section 2.1 of this Unit applies here, for we may regard the interval T as being the time period characteristic of the electromagnetic wave as indicated in Figure 7. As O shines his signal out towards O' he (O) sees the wave rise and fall regularly with the period T.

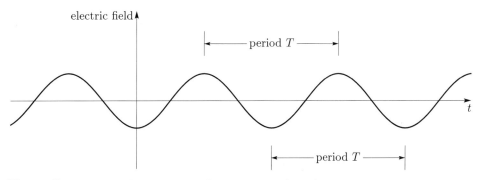

Figure 7   The period T corresponds to one wavelength.

But the period that O′ measures will be $T' = k(V)\,T$. Since the period T and frequency of the light are related by the formula $T = 1/f$, we have:

$$f' = \frac{1}{T'} = \frac{1}{k(V)\,T} = \frac{f}{k(V)}$$

or (using Equation 4):

$$f' = \frac{f}{k(V)} = f\sqrt{\frac{1 - \dfrac{V}{c}}{1 + \dfrac{V}{c}}}. \tag{5}$$

Let us see if this result bears out our expectations.

Objectives 2 and 7   **SAQ 7**   Show that if O and O′ recede from each other then O′ measures a frequency f' which is less than f (i.e. $f' < f$ for $V > 0$), and vice versa if O and O′ approach one another (i.e. $f' > f$ for $V < 0$).

Does Equation 5 make sense when O and O′ are stationary with respect to each other (i.e. $V = 0$)?

Objectives 2 and 7   **SAQ 8**   Make a rough sketch of f' as a function of V/c for all values of V/c between -1 and $+1$.

2.3   A derivation of the Lorentz transformation

Study comment

In your study of Section 2.3, we expect you to be able to follow the algebraic manipulations in our derivation of the Lorentz transformation, but we do *not* require you to be able to reproduce the full mathematical argument from memory. The line of argument using spacetime diagrams is, however, basic to understanding relativity, and you will find that such diagrams permeate the Block. Although the mathematical *derivation* of the Lorentz transformation (Equations 23a–e) is not examinable (see the wording of Objective 1), we feel that its physical meaning *is* examinable; it is of paramount importance to the Block and much of the rest of the Course.

We now consider further the connection between the coordinates (t, x) and (t', x') assigned to an event $\mathscr{E}$ by the observers O and O′ of Figure 2. As before we assume, for simplicity, that O and O′ set their clocks to zero when their origins pass one another. In other words, the special event at $x = 0$, $t = 0$ corresponds to $x' = 0$, $t' = 0$, and vice versa. This can always be done (in a thought-experiment that explores the consequences of a logically conceivable situation) and no physical results are lost by doing so.

Consider then any event $\mathscr{E}$ as shown in Figure 8. For the time being let us assume that $\mathscr{E}$ occurs somewhere along the collinear x- and x'-axes of the frames as depicted in Figure 2 (i.e. $x^2 = x^3 = 0$). We may suppose also

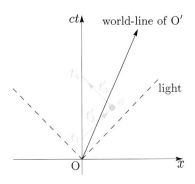

Figure 8 An arbitrary event $\mathscr{E}$ at point (t, x) as seen by O and point (t', x') as seen by O'. Points labelled t_1, t_4 etc. correspond to coordinates ct_1, ct_4 etc.

that $\mathscr{E}$ is triggered by a pulse emitted at time t_1 on the clock of O, at his origin $x = 0$, and received back, at $x = 0$, at time t_4. The pulse will be measured by O' to cross his origin ($x' = 0$) at time t'_2, say, on its way out and to recross at time t'_3, say, on its way back.

Observer O will assign time $t = \frac{1}{2}(t_4 + t_1)$ and position $x = \frac{1}{2}c(t_4 - t_1)$ to event $\mathscr{E}$. Observer O' can think of $\mathscr{E}$ as triggered by *him* by a pulse sent out at his time t'_2 and received back at his time t'_3. Thus O' will assign time $t' = \frac{1}{2}(t'_3 + t'_2)$ and position $x' = \frac{1}{2}c(t'_3 - t'_2)$ to $\mathscr{E}$. To summarize so far, we have for the coordinates of $\mathscr{E}$ in the two frames,

$$(t, x) = \left(\frac{(t_4 + t_1)}{2}, \frac{c(t_4 - t_1)}{2} \right), \tag{6}$$

$$(t', x') = \left(\frac{(t'_3 + t'_2)}{2}, \frac{c(t'_3 - t'_2)}{2} \right). \tag{7}$$

Note that Equations 6 and 7 together are actually *four* equations. From (6), for example, we can equate t with $\frac{1}{2}(t_4 + t_1)$ and x with $c(t_4 - t_1)/2$.

Equations 6 and 7 relate the four quantities x, t, x' and t' to the four quantities t_1, t_4, t'_2 and t'_3, but what we want is a relationship between x, t, x' and t'. We must therefore eliminate t_1, t_4, t'_2 and t'_3 from the two equations. To do this we need *further* relationships between these variables.

From our analysis of Section 2.2 we can write two further equations, for we know that the time interval $t_1 - 0 = t_1$ on the clock of O at his origin is related to the time interval $t'_2 - 0 = t'_2$ by

$$t'_2 = k(V)\, t_1. \tag{8}$$

This follows since we can think of O sending out pulses at times $t = 0$ and $t = t_1$ by his clock. The first pulse would be 'received' instantaneously by O' at time $t = 0$ (since O and O' are together at that time); the second would be received by O' at time $t = t'_2$ on his clock. Thus t'_2 is the interval, on his clock, between the reception of these two pulses by O'.

In the same way the time intervals $t'_3 - 0 = t'_3$ and $t_4 - 0 = t_4$ are related by:

$$t_4 = k(V)\, t'_3. \tag{9}$$

Equations 6, 7, 8 and 9 do the job. To see this most efficiently (we think!) consider the following steps:

Step 1

From (6) $ct + x = \dfrac{c(t_4 + t_1)}{2} + \dfrac{c(t_4 - t_1)}{2} = ct_4. \tag{10}$

From (7) $ct' + x' = \dfrac{c(t'_3 + t'_2)}{2} + \dfrac{c(t'_3 - t'_2)}{2} = ct'_3. \tag{11}$

Substituting for t_4 from (9) in (10):

$$ct + x = ct_4 = ck(V)t'_3. \tag{12}$$

Substituting for ct'_3 from (11) in (12):

$$ct + x = k(V)(ct' + x'). \tag{13}$$

Step 2

From (6) $\quad ct - x = \dfrac{c(t_4 + t_1)}{2} - \dfrac{c(t_4 - t_1)}{2} = ct_1.$ $\hfill (14)$

From (7) $\quad ct' - x' = \dfrac{c(t_3' + t_2')}{2} - \dfrac{c(t_3' - t_2')}{2} = ct_2'.$ $\hfill (15)$

Substituting for t_2' from (8) in (15):

$$ct' - x' = k(V)\,ct_1, \hfill (16)$$

and substituting for ct_1 from (14) in (16):

$$ct' - x' = k(V)(ct - x). \hfill (17)$$

In this way, with Equations 13 and 17 we have achieved our goal of relating the coordinates (t, x) of $\mathscr{E}$ in one inertial system to the coordinates (t', x') of $\mathscr{E}$ in the other inertial system. (Remember $k(V)$ is known, and given by Equation 4 in Section 2.1.) But there is a more customary way of expressing this Lorentz transformation.

Objective 1   **SAQ 9**   Carry out the algebra to show from Equations 4, 13 and 17 that the Lorentz transformation can also be written

$$t' = \gamma(V)\left(t - \frac{V}{c^2}x\right) \hfill (18)$$

$$x' = \gamma(V)(x - Vt) \hfill (19)$$

where we define

$$\gamma(V) = \frac{1}{\sqrt{1 - \left(\dfrac{V}{c}\right)^2}}. \hfill (20)$$

Lorentz transformation

Equations 18, 19 and 20 comprise the usual form of the *Lorentz transformation* connecting the coordinates of an event on the collinear x-axes as measured by the two observers of Figure 2. Let us stand back a little from these results to see if they make good sense.

In the first place, let us consider what the Lorentz transformation becomes in the limit where $|V| \ll c$. This is known as the *non-relativistic* or *Galilean limit*; after all, in our everyday world, most things do move with $|V| \ll c$. So now we must compare the Lorentz transformation, Equations 18, 19 and 20, to the Galilean transformation of SAQ 1 in Section 2.1, and the following important SAQs lead you through this.

Objective 1   **SAQ 10**   Show that in the non-relativistic limit, Equations 20 and 19 reduce to

$$\gamma(V) = 1$$
$$x' = x - Vt.$$

Objective 1   **SAQ 11**   Show that when V/c is so small as to be negligible, Equation 18 becomes $t' = t$.

So in the limit where $|V|/c \longrightarrow 0$, $t = t'$ and time is seemingly absolute: all observers of any event would agree on the time at which it occurred. In the solved example in Section 4.3 of Unit 5, which was concerned with a method of synchronizing clocks in any given inertial frame (described in the same Section), it was shown that in the mathematical limit when c is infinite all the field of synchronized clocks are *seen* by an observer in that frame to read the same time. Of course c is *not* infinite, but as far as the Lorentz transformation equations are concerned, $|V| \ll c$ amounts to the same as having large c if V is known to be of everyday magnitude. Therefore, if the *ratio* $|v|/c$ is negligibly small, where v is any relevant physical velocity, the relativistic results will be close to the pre-Einstein prediction.

In the second place, when $|V|/c$ is *not* small, the Lorentz transformation differs *drastically* from the Galilean situation.

Objective 1   SAQ 12   Calculate $\gamma(V)$ for $|V| = 0.01c, 0.1c, 0.3c, 0.95c, 0.995c$. Assuming that V can never exceed the speed of light c, sketch a rough graph of $\gamma(V)$ versus V/c. (The electrons in a colour TV tube reach $|V| \sim 0.3c$.)

When $|V|/c$ is not small, i.e. in the *relativistic* domain, a new feature of relativity comes into play: space and time become 'mixed up'. To see this, notice that in Equation 18, t' is a linear combination of t and x. This feature will be central to much of what follows in the Course, and indeed is already reflected in our use of the term *spacetime*. It was Minkowski in 1908 who first emphasized the consequences of mixing t and x. The collection of all events (ct, x^1, x^2, x^3) is sometimes referred to as *Minkowski spacetime*.

In the third place, the theory of special relativity, postulate SR1, states that all inertial frames are equivalent for the observation of physics. In other words, no one inertial frame is special for taking measurements. In the present context we can say that if (as in Figure 2) O sees O' moving off to the right (along the positive x-axis) with velocity V, then O' must see O moving off to the left (along the negative x'-axis) with velocity $-V$, and that the Lorentz transformation had better draw no other distinction between the two frames. This is reflected in the inverse transformation, which you will now discover:

Objective 1   SAQ 13   Show by solving Equations 18, 19 and 20 algebraically for x and t in terms of x' and t' that, for the situation depicted in Figure 2,

$$x = \gamma(V)(x' + Vt') \tag{21}$$

$$t = \gamma(V)\left(t' + \frac{V}{c^2}x'\right). \tag{22}$$

Equations 18, 19 and 20 give the coordinates (t', x') assigned by O' to an event $\mathscr{E}$ in terms of the coordinates (t, x) assigned by O to the same event. The answer to SAQ 13 shows that we may *reverse* this process by simply replacing V by $-V$ and interchanging x with x' and t with t'.

To complete this Section on the Lorentz transformation, we shall find the coordinates assigned by the two observers O and O′ of Figure 1 to a perfectly arbitrary event $\mathscr{E}$ which not only occurs away from the origin but also occurs *off* the x^1- and x'^1-axes. The coordinates assigned to the event by O are (t, x^1, x^2, x^3) and those assigned by O′ are (t', x'^1, x'^2, x'^3). It turns out from very general arguments that the full transformation equations are:

$$t' = \gamma(V)\left(t - \frac{V}{c^2}x^1\right) \tag{23a}$$

$$x'^1 = \gamma(V)(x^1 - Vt) \tag{23b}$$

$$x'^2 = x^2 \tag{23c}$$

$$x'^3 = x^3 \tag{23d}$$

$$\gamma(V) = \frac{1}{\sqrt{1 - \left(\dfrac{V}{c}\right)^2}}. \tag{23e}$$

The key elements in the arguments are (i) that the transformation equations are linear, i.e. of form $x'^2 = a + bx^1 + cx^2 + dx^3 + et$, and, (ii) the fact that in Figure 1, $x^2 = 0$ and $x'^2 = 0$ coincide permanently as do $x^3 = 0$ and $x'^3 = 0$. The requirement for linearity follows from SR1 and the fact that linear motion in one frame becomes linear motion in the other. Also, of course, the transformation must be such as to reduce to the Galilean form when $|V|/c$ is small. An alternative argument will be given in Section 3.2 of Unit 7.

Objective 1   SAQ 14   (a) Say in words what Equations 23a to 23e describe. One or two sentences will suffice.

(b) Write down the inverse transformation (cf. SAQ 13).

(c) Write out Equations 23a and b in terms of variables x^1, x'^1, ct, ct'.

3 Applications of the Lorentz transformation

Now that we have the Lorentz transformation, space and time become permanently entangled; we consider just a few examples.

3.1 Length contraction

Suppose a rod moves at constant speed V to the right along the x^1-axis of an observer O. For simplicity, let us suppose that the rod is aligned parallel to the x^1-axis. The world-lines of the two ends of the rod according to O are shown in Figure 9. What is the length of the rod according to observer O? (Note: we observe the convention introduced in Unit 5 whereby we use x for x^1 where there is no chance of confusion.)

First let us note that since the rod moves uniformly there is certainly an inertial frame in which the rod is stationary. Such a frame is called a *rest frame* of the rod. A rest frame will be any inertial frame that moves to the right at speed V. Let us call the length measured in such a rest frame the *rest length* of the rod, and in the present case denote it by l_0.

rest frame

rest length

Now, as has been discussed for example in Section 4.2 of Unit 5, it is especially important in thought-experiments to specify exactly *how* a measurement is to be made. Perhaps the most systematic measurement for O to make is to locate the two ends of the rod at what is in *his* frame the *same time*.

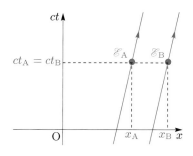

The situation is depicted in Figure 9. $\mathscr{E}_A$ is the event that the rear end of the rod is found by O to be located at the point x_A at time t_A; similarly, the coordinates of the front end of the rod at $\mathscr{E}_B$ are t_B and x_B. We have $t_A = t_B$ since O locates events $\mathscr{E}_A$ and $\mathscr{E}_B$ at the same time.

Figure 9 If the rod is of length l_0 according to observer O′ for whom the rod is at rest, what is its observed length according to O?

The rest frame of the rod most convenient to our present analysis will be a frame O′ with its x'^1-axis running along x^1, as in Figure 1. Then the rod will be aligned parallel to the x'^1-axis. The frame O′ moves along to the right with velocity V. The coordinates of event $\mathscr{E}_A$ as measured by O′ are (t'_A, x'_A), while those of $\mathscr{E}_B$ are (t'_B, x'_B). Since by now we know that simultaneity is a relative concept (see Section 4.2 of Unit 5) we *cannot* say that $t'_A = t'_B$.

We can, however, apply the Lorentz transformation, Equations 23, to connect the coordinates of $\mathscr{E}_A$ and $\mathscr{E}_B$ as measured by observers O and O′. From Equation 23b we have:

$$x'_B = \gamma(V)(x_B - V t_B)$$
$$x'_A = \gamma(V)(x_A - V t_A).$$

As we have said, observer O is measuring the location of both ends of the rod at the same time, so that $t_A = t_B$. Thus, subtracting the second equation above from the first gives

$$x'_B - x'_A = \gamma(V)(x_B - x_A)$$

or

$$x_B - x_A = \frac{1}{\gamma(V)}(x'_B - x'_A).$$

This is the desired result; it relates the length, $(x_B - x_A)$, of the rod according to O to the rest length, $(x'_B - x'_A)$. Defining l as $(x_B - x_A)$ and

length contraction formula

identifying l_0 as $(x'_B - x'_A)$ and using Equation 20 for $\gamma(V)$ gives the famous *length contraction formula*:

$$l = l_0 \sqrt{1 - \left(\frac{V}{c}\right)^2} = \frac{l_0}{\gamma(V)}. \qquad (24)$$

This effect should *not* be construed as a mechanical contraction of the rod, but a consequence of the properties of spacetime.

Objectives 1 and 3   SAQ 15   (a) Say briefly in words what Equation 24 describes mathematically.

(b) What does Equation 24 reduce to in the non-relativistic limit $V/c \longrightarrow 0$ and in the extreme relativistic limits of $V/c \longrightarrow +1$ and $V/c \longrightarrow -1$?

Objectives 1 and 3   SAQ 16   Suppose the rod, of rest length l_0, moves as before to the right at speed V, but that this time it is aligned perpendicular to the x^1-axis, say along x^2. What will the length l be, as measured by O? (*Hint*: Refer to Equations 23c and 23d.)

3.2   Time dilation

Another famous effect in special relativity is the slowing down of moving clocks. This has to be taken into account, for example, when calculating the distance travelled by the unstable fundamental particles created by cosmic rays when they arrive in the Earth's upper atmosphere from outer space.

To simplify things, consider a clock moving uniformly to the right, say, at velocity V in frame O along the x-axis. The clock's world-line is shown in Figure 10. As before, we can introduce a rest frame, O', for the clock. This moves to the right with the clock at velocity V. We shall consider the time interval between two ticks of the clock, as measured by O and by O'.

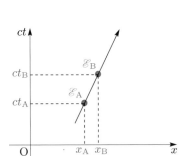

Figure 10   The clock's world-line according to O.

Let us denote the event of one tick by $\mathscr{E}_A$ and of the next tick by $\mathscr{E}_B$. Then we require a relationship between the interval in time, $(t'_B - t'_A)$, from one tick to the next as measured in the rest frame O', and the time interval, $(t_B - t_A)$, between the same two ticks as measured by O. The relevant equation is 23a. It gives:

$$t'_B = \gamma(V) \left(t_B - \frac{V}{c^2} x_B \right)$$

$$t'_A = \gamma(V) \left(t_A - \frac{V}{c^2} x_A \right).$$

Subtracting the second equation from the first gives:

$$t'_B - t'_A = \gamma(V) \left[(t_B - t_A) - \frac{V}{c^2}(x_B - x_A) \right]. \qquad (25)$$

According to O, the clock is moving with velocity V. Therefore we can say:

$$x_B - x_A = (t_B - t_A)V.$$

Substituting for $(x_B - x_A)$ in Equation 25 gives:

$$t'_B - t'_A = \gamma(V) \left[(t_B - t_A) \left(1 - \frac{V^2}{c^2} \right) \right].$$

Substituting for $\gamma(V)$, from Equation 23e, gives:

$$t'_B - t'_A = (t_B - t_A) \sqrt{1 - \frac{V^2}{c^2}} = \frac{1}{\gamma(V)} (t_B - t_A).$$

Rearranging,

$$t_B - t_A = \frac{t'_B - t'_A}{\sqrt{1 - \dfrac{V^2}{c^2}}} = \gamma(V)(t'_B - t'_A).$$

Finally, if we denote by τ_0 the interval between two ticks in the rest frame, and by τ the corresponding time interval in frame O, we have the famous *time dilation formula*:

time dilation formula

$$\tau = \tau_0 \frac{1}{\sqrt{1 - \left(\dfrac{V}{c} \right)^2}} = \gamma(V)\tau_0. \qquad (26)$$

This is perhaps the most famous result of the 'relativity of time'.

We certainly advise you to run over in your mind the steps leading to Equation 26. Particularly notice the contrast between Equations 24 and 26. For Equation 24 says (briefly) that a moving metre stick appears shortened by a factor $1/\gamma(V)$, while Equation 26 says that the interval between ticks of a moving clock appear to *increase*, i.e. a moving clock appears to tick more slowly by a factor $\gamma(V)$, since $\gamma(V) \geqslant 1$ for all values of V lying between plus and minus c.

A practical application of Equation 26 is the aforementioned case of unstable fundamental particles moving at speeds near the speed of light down towards the Earth, after being created high up in the Earth's atmosphere by cosmic rays from outer space. This is indicated schematically in Figure 11. Typically, event $\mathscr{E}_A$ might be the creation of a muon. This particle has a mass of about 207 times that of an electron and is unstable. It decays after an average time of about 2.2×10^{-6} s (in its rest frame), splitting up into an electron and two other fundamental particles called neutrinos. This decay is indicated as event $\mathscr{E}_B$ in Figure 11. A land-based experiment on muon lifetime is discussed in Section 5.3 of Unit 7.

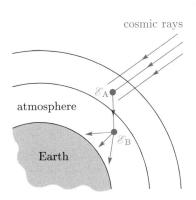

cosmic rays

atmosphere

Earth

Figure 11 A cosmic ray particle interacts with an atom in the upper atmosphere, event $\mathscr{E}_A$ and subsequently decays, event $\mathscr{E}_B$.

Objective 4

SAQ 17   Consider the creation of a muon, event $\mathscr{E}_A$ of Figure 11, and its subsequent decay, event $\mathscr{E}_B$. Suppose a muon, which decays *in its rest frame* after about 2×10^{-6} s, moves at a speed of $0.95c$ (which is typical). Find the decay time as measured by an observer fixed to the Earth. How far (travelling in a straight line) would the muon move, at this speed, between creation and decay, according to this observer? How does this compare with the distance calculated as if *non*-relativistic physics applied, namely $(2 \times 10^{-6} \text{ s}) \times (0.95) \times (3 \times 10^8 \text{ m s}^{-1}) = 5.7 \times 10^2 \text{ m}$?

Physicists dealing with unstable fundamental particles apply the time dilation formula throughout their work; indeed, time dilation makes it practical to do experiments with certain particles of fleeting existence.

Objectives 1 and 4   SAQ 18   Apply the inverse transformation of Equation 23a to give an alternative derivation of Equation 26.

3.3   The transformation of velocities

Having dealt with the transformation of coordinates and seen a few applications, we now turn to the question of relating the velocity measurements of different inertial observers.

Suppose we have a particle moving in the positive x^1-direction of an inertial frame as in Figure 12. For simplicity, we take the x^2- and x^3-components of the velocity to be zero in O.

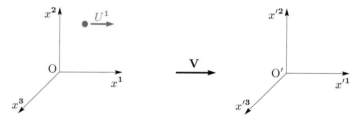

Figure 12   The body moves in the x^1- (and x'^1-) direction.

In the spacetime diagram for O, the world-line of the particle will be as in Figure 13. Let us denote by U^1 the particle's velocity along x^1 in frame O (we often use x for x^1 when there is no chance of confusion.) We want to find the velocity along x'^1 of the body as measured by some other inertial observer O', as in Figure 12. Denote this velocity by U'^1.

To find U'^1, consider two infinitesimally separated points (events) $\mathscr{E}_A$ and $\mathscr{E}_B$ on the particle's world-line. Let us denote by Δx the separation $(x_B - x_A)$, where x_B and x_A are the positional coordinates of $\mathscr{E}_B$ and $\mathscr{E}_A$ in system O. Similarly, $\Delta t = t_B - t_A$. Then, to observer O, $U^1 = \Delta x/\Delta t$ in the limit $\Delta t \longrightarrow 0$. To find U'^1 we need merely find $\Delta x' = x'_B - x'_A$ and $\Delta t' = t'_B - t'_A$, where (t'_B, x'_B) and (t'_A, x'_A) are the coordinates of $\mathscr{E}_B$ and $\mathscr{E}_A$ in the frame O'.

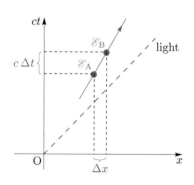

Figure 13   The body's world-line according to O.

These follow easily from the Lorentz transformation, Equations 23. For instance, from Equation 23a we have:

$$t'_B = \gamma(V)\left(t_B - \frac{V}{c^2}x_B\right),$$

$$t'_A = \gamma(V)\left(t_A - \frac{V}{c^2}x_A\right).$$

Subtracting the second equation from the first gives:

$$t'_B - t'_A = \Delta t' = \gamma(V)\left[(t_B - t_A) - \frac{V}{c^2}(x_B - x_A)\right],$$

$$\Delta t' = \gamma(V)\left(\Delta t - \frac{V}{c^2}\Delta x\right). \tag{27}$$

With a little hindsight this result is obvious; it simply says that time differences (like Δt and $\Delta t'$) and position differences (such as Δx and

$\Delta x'$), separating any two not necessarily adjacent events in spacetime, transform in the usual way. Clearly, from Equation 23b we shall also have:

$$\Delta x' = \gamma(V)(\Delta x - V\Delta t). \tag{28}$$

Finally, in our simplified case, the particle is not moving in the x^2- or x^3-directions, so $\Delta x^2 = x_B^2 - x_A^2 = 0$ and $\Delta x^3 = x_B^3 - x_A^3 = 0$. Thus, from Equations 23 we can say

$$\Delta x'^3 = \Delta x^3 = 0 \qquad \text{and} \qquad \Delta x'^2 = \Delta x^2 = 0.$$

The velocity of the particle follows from Equations 27 and 28. We have:

$$U'^1 = \frac{\Delta x'}{\Delta t'}$$

$$= \frac{\Delta x - V\Delta t}{\Delta t - \dfrac{V}{c^2}\Delta x} \qquad \text{(cancelling } \gamma(V))$$

$$= \frac{\dfrac{\Delta x}{\Delta t} - V}{1 - \dfrac{V}{c^2}\dfrac{\Delta x}{\Delta t}} \qquad \text{(dividing above and below by } \Delta t)$$

$$\boxed{U'^1 = \frac{U^1 - V}{1 - \dfrac{V}{c^2}U^1}}. \tag{29}$$

Before proceeding, we suggest that you verify that this expression makes good sense by attempting the following self-assessment question.

Objective 5   SAQ 19 (a) Say briefly what Equation 29 describes. What do the terms in this equation represent?

(b) What does Equation 29 give in the non-relativistic limit $\dfrac{U^1 V}{c^2} \longrightarrow 0$? Does this result make sense?

(c) Equation 29 is written to give U'^1 in terms of U^1. Invert the equation to give U^1 in terms of U'^1. With a little thought you might be able simply to write down the answer.

We are now in a position to show that no material body that moves at a speed less than c in one inertial frame can appear to move at a speed equal to or exceeding c in any other inertial frame. For reconsider Figure 12. Let us assume that the particle moves in the positive x'^1-direction in frame O$'$ at some speed U'^1, where $c > U'^1 > 0$. And suppose O$'$ moves to the right relative to O at some speed V less than c, so that $c > V > 0$. Then clearly anyone attached to axes x^1, x^2 and x^3 will see the particle move to the right, along the positive x^1-axis, at some speed U^1 greater than U'^1. From the solution to SAQ 19 we have:

$$U^1 = \frac{U'^1 + V}{1 + \dfrac{U'^1 V}{c^2}}. \tag{30}$$

In our present application we are considering that U'^1 and V are positive, but that each is less than c. For instance, suppose that $U'^1 = 0.9c$ and $V = 0.9c$. Then if non-relativistic mechanics applied we would have $U^1 = 1.8c$. But Equation 30 gives:

$$U^1 = \frac{1.8}{1 + 0.81}c = \frac{1.8}{1.81}c = (0.994\,47\ldots)c.$$

In general, if U'^1 and V are less than c you ought to be able to convince yourself that Equation 30 always gives a value for U^1 that is also less than c. Now, if we introduce a *third* frame moving at some velocity less than c with respect to O', it will still be impossible to produce a velocity exceeding c. Clearly, the introduction of any number of new frames, each moving at any speed less than c with respect to its predecessor, will fail to show a particle moving at a speed exceeding c.

Objective 5

SAQ 20   Now suppose the particle has components of velocity U^2 and U^3 in the x^2- and x^3-directions respectively of frame O. Show that the corresponding x'^2- and x'^3- components of velocity in the frame O' of Figure 12 are given by

$$U'^2 = \frac{U^2}{\gamma(V)\left(1 - \dfrac{U^1 V}{c^2}\right)} \tag{31}$$

$$U'^3 = \frac{U^3}{\gamma(V)\left(1 - \dfrac{U^1 V}{c^2}\right)}. \tag{32}$$

Objective 5

SAQ 21   What would Equation 30 give for the case of $U'^1 = c$? How would you interpret your result?

3.4   Charge invariance: The thought-experiment of Section 2 of Unit 5 revisited

We now know enough to show that the line of reasoning followed in our thought experiment of Section 2 of Unit 5 was consistent. You will recall that it seemed that length measurements are relative to the (inertial) observer. This should now seem plausible from Equation 24 in Section 3.1.

However, there was one point of worry; in writing Equations 2a and 2b (Lorentz force law) in Unit 5 we assumed that if a particle has charge q according to one inertial observer then it has the same charge q for any other inertial observer. In other words, we supposed that charge is a Lorentz invariant, unaffected by the Lorentz transformation just as is the speed of light, c. However, we deduced that a single line of steady current has a different charge per unit length for two different inertial observers. For, accepting the experimental fact that the wire is observed to have zero net charge per unit length in its rest frame, we concluded that it has a net positive charge per unit length in an inertial frame moving uniformly along with the velocity of the current-carrying electrons. Is then, our assumption that charge is a Lorentz invariant inconsistent? Have we not created a net charge by going to the new frame? The solution to this paradox lies in the experimental fact that steady currents only flow in *closed* circuits. Thus, to be realistic, we ought to reject Figure 14(a) in favour of Figure 14(b). We have simplified Figure 14(b) by leaving out the source needed to keep current flowing. This is permissible since we are considering an idealized thought-experiment — and anyway it is possible nowadays to bring about the situation shown in Figure 14(b) by using a superconducting circuit.

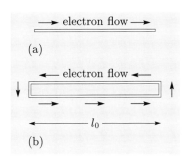

(a)

(b)

Figure 14   Current flow in the rest frame of the wire (the laboratory frame).

To be specific, let us suppose the length of the circuit in the laboratory (rest) frame is l_0, as indicated in Figure 14(b). Reverting to the style of the Figures in Section 2 of Unit 5, we then have the situation depicted in Figure 15(a) for the frame lying at rest with respect to the wire (the rest frame). The current-carrying electrons move with a steady velocity $\pm\mathbf{u}$. For convenience, we have labelled the two identical wires in Figure 15(a) as wire 'a' and wire 'b'.

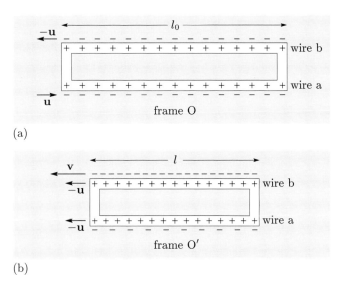

(a)

(b)

Figure 15   (a) In frame O, the charge per unit length in each wire for electrons and positive ions is equal and opposite. (b) What is $\mathbf{V}$, the velocity of the electrons in wire 'b' as observed in O'?

The net charge per unit length in rest frame O is known from experiment to be zero for wire 'a' and wire 'b' separately. Thus, if the number of positive and negative (electron) charges per unit length for wire 'a' are λ_a^+ and λ_a^-, respectively, then $\lambda_a^+ = \lambda_a^-$ in frame O. Similarly, for wire 'b' we must have $\lambda_b^+ = \lambda_b^-$ in frame O. Furthermore, both wires are identical and at rest in the laboratory frame O, so we have $\lambda_a^+ = \lambda_b^+$. Let us define the number of positive charges per unit length in frame O as λ_0. Then we have:

$$\lambda_a^+ = \lambda_b^+ = \lambda_a^- = \lambda_b^- \equiv \lambda_0. \tag{33}$$

Things get more interesting when we observe the circuit from a frame moving uniformly along to the right, relative to O, with velocity $\mathbf{u}$ (Figure 15(b)). Label this frame O'. In O' the electrons of wire 'a' lie at rest. Furthermore, since the positive charges of wires a and b lie at rest in O they will move to the left in O' with velocity $-\mathbf{u}$. Also, whereas the electrons of wire 'b' drift to the left in O with velocity $-\mathbf{u}$, in frame O' they will move *faster* to the left with velocity $\mathbf{v}$, say.

Objectives 5 and 6   SAQ 22   Show that $v = |\mathbf{v}|$ is given by the expression

$$v = \frac{-2u}{1 + \left(\dfrac{u}{c}\right)^2}. \tag{34}$$

Now in Section 2 of Unit 5 we argued that, as observed by O', wire 'a' will have a *positive* charge per unit length. Figure 15(b) has been drawn to indicate this. Let us now prove it.

Let the number of positive charges per unit length in wire 'a' as observed by O' be denoted by $\lambda_a'^+$. These positive charges are distributed uniformly along wire 'a' (we ignore the end effects and suppose the wire to be long), and are observed by O' to move uniformly to the left with velocity $-\mathbf{u}$. Thus if O' observes each wire to have length l, say, moving to the left with velocity $-\mathbf{u}$, then the *number* of positive charges encompassed by l will be $\lambda_a'^+ l$. But, as observed by O, this same length will encompass the same number of positive charges. In other words, we are assuming that the number of positive charges encompassed by this length is the same for all inertial observers, i.e. is Lorentz invariant.

Now the number of positive charges per unit length is just λ_0 in the frame of O. But, as observed by O, the length will lie at rest since the positive charges lie at rest, and so will have its rest value, l_0 say. The number of positive charges encompassed is the same for O as for O'.

Thus

$$\lambda_a'^+ l = \lambda_0 l_0.$$

We may apply Equation 24 here to relate l and l_0. Since the length moves at velocity $-\mathbf{u}$ according to O', we have:

$$l = l_0 \sqrt{1 - \left(\frac{u}{c}\right)^2}.$$

The preceding two equations give

$$\lambda_a'^+ = \frac{\lambda_0}{\sqrt{1 - \left(\dfrac{u}{c}\right)^2}}. \tag{35}$$

Objective 6   SAQ 23   Show briefly that $\lambda_a'^+$ is greater than λ_0, as anticipated in SAQs 4 and 5 of Unit 5.

We shall now find $\lambda_a'^-$, the number of negative charges per unit length in wire 'a' as observed by O'. These negative charges are at rest relative to O' whereas they are moving relative to O with speed u. We shall therefore get a similar relationship for the density of negative charge (in wire 'a') as we had for the density of positive charge (in wire 'a'). All we need to do is replace in Equation 35 the symbol for stationary positive charges (λ_0) by that for stationary electrons ($\lambda_a'^{-}$), and the symbol for moving positive charges ($\lambda_a'^{+}$) by that for moving electrons (λ_0):

$$\lambda_0 = \frac{\lambda_a'^{-}}{\sqrt{1 - \left(\dfrac{u}{c}\right)^2}}.$$ (36)

According to Equations 35 and 36 the density of electrons per unit length is observed by O′ to be *less* than the density of positive charges. This was what we had anticipated. We have shown that, according to O′, wire 'a' has a net positive charge. This is very satisfactory in as much as it reconciles the views of the two observers O and O′ with regard to the forces exerted by the current-carrying wire. For you may recall that the upshot of Section 2 of Unit 5 was that one observer's electric field may be another observer's magnetic field.

But we are left with a problem: if the *density* of charges has changed, what has happened to their *number*? If we are to retain electric charge invariance, both O and O′ must agree on the total number of charges.

As far as the positive charges are concerned there is no difficulty. Their *density* went up simply because the length of wire was reduced. Observer O′ agrees with O that the number of positive charges on wire 'a' (and also on wire 'b') is $\lambda_0 l_0$.

But what of the negative charges? While the length of the wire was reduced (as observed by O′), the density of electrons was *also* reduced to a value below λ_0 (see Equation 36). Therefore, the number of electrons in wire 'a' according to O′ is less than that according to O. To be specific, letting n_a' be the number of electrons in wire 'a' according to O′, we have

$$\begin{aligned} n_a' &= l\lambda_a'^{-} \\ &= l_0\sqrt{1 - \frac{u^2}{c^2}} \times \lambda_0\sqrt{1 - \frac{u^2}{c^2}} \\ &= \lambda_0 l_0 \left(1 - \frac{u^2}{c^2}\right). \end{aligned}$$ (37)

This is less than the number of electrons $\lambda_0 l_0$, observed by O. Where have the other electrons gone?

To answer this question we must consider what is happening in the return wire, wire 'b'. Neither frame O nor frame O′ is a rest frame for the negative charges in this other wire (see Figures 15(a) and (b)). We can, however, relate the density of electrons in wire 'b' as seen by O′, namely $\lambda_b'^{-}$, to the density of the stationary electrons in wire 'a' as seen by O′, namely $\lambda_a'^{-}$. The density of electrons in wire 'b', as seen by O′, is given by an equation similar to Equation 35, where the velocity v of the electrons in wire 'b' replaces the former velocity u, $\lambda_b'^{-}$ replaces $\lambda_a'^{+}$, and $\lambda_a'^{-}$ replaces λ_0. Thus we have (see Figure 15(b)):

$$\lambda_b'^{-} = \frac{\lambda_a'^{-}}{\sqrt{1 - \left(\dfrac{v}{c}\right)^2}}.$$ (38)

Objective 6   SAQ 24   Show that

$$\lambda_{\mathrm{b}}'^{-} = \lambda_0 \frac{1 + \left(\dfrac{u}{c}\right)^2}{\sqrt{1 - \left(\dfrac{u}{c}\right)^2}}. \tag{39}$$

[Hint: you will need to use Equation 34.]

We may now compute the total number of electrons in wire 'b' according to O'. Since the length of wire 'b' is the same as that of wire 'a', letting n_{b}' be the number of electrons in wire 'b' according to O', we have (using Equation 39)

$$n_{\mathrm{b}}' = l\lambda_{\mathrm{b}}'^{-} = \lambda_0 l_0 \left(1 + \frac{u^2}{c^2}\right). \tag{40}$$

This compares with n_{a}', the number of electrons observer O' attributes to wire 'a', which we saw in Equation 37 was $\lambda_0 l_0 (1 - u^2/c^2)$. Adding n_{a}' to n_{b}' we have:

total number of electrons according to O'

$$= \lambda_0 l_0 \left(1 - \left(\frac{u}{c}\right)^2\right) + \lambda_0 l_0 \left(1 + \left(\frac{u}{c}\right)^2\right) = 2\lambda_0 l_0.$$

This is the same total number as observer O attributes to the two wires. So the answer to the question of where the missing electrons had gone, is that they were on wire 'b' instead of wire 'a'! This raises an interesting point to do with the relativity of the simultaneity of events separated in space, as brought out in the following SAQ.

Objective 6   SAQ 25   Using this last remark as a hint, explain why O and O' cannot simply count the number of electrons on each wire and agree as to their number. Then explain why they *are* able to count the number of positive charges on each wire and agree on the result.

Objective 6   SAQ 26   What did we set out to prove in this Section? Did we succeed?

Thus, in Section 3.4, we have shown how subtle the effect of the relativity of simultaneity can be.

4 'Paradoxes' in special relativity

Physicists who devote their time to studying the subatomic world of fundamental particle physics use the special theory in every step of their experimenting and theorizing. In their experiments, the speeds of their particles are usually close to c; indeed, the vast accelerators like those at CERN would not work if constructed to Newtonian dynamics. The theory of fundamental particles depends throughout on special relativity; we shall return to this in the next Unit.

Granting, then, that we accept the overwhelming confirmation of special relativity, most people still find themselves sometimes struggling to reconcile apparent paradoxes. The main problem is that it is difficult to accept that time actually is relative and stands, in a sense, on an equal footing with space. You have seen, however, that special relativity insists that different inertial observers will not agree on the simultaneity of two events and that the time and position coordinates of any event are 'mixed up' together *differently* for *different* inertial observers according to the Lorentz transformation, Equations 23a to 23e.

Another source of trouble is that one often forgets that in any real or thought experiment it is important to specify *how* each measurement is to be made. It seems to come back to the concept of 'observer', discussed in Section 6 of Unit 5.

Consider the following conundrums as examples. These are not real paradoxes, since it is impossible to construct a true paradox from a well-defined physical theory. However, thinking about them is a good way of getting to grips with the theory.

4.1 Is Doppler shift consistent with time dilation?

In Section 3.2 we derived the famous time dilation formula, Equation 26. This equation is often described as saying that 'moving clocks run slow'. But what does this mean? Does it mean that if you look (actually look with your eyes, or shoot a film) at a clock approaching you, you will see it run slow compared with an identical clock held in your hand?

The answer is no. In fact you will *see* with your *eyes* that the clock runs *faster* as it approaches you and runs slower as it leaves you. Then what does the statement 'moving clocks run slow' mean?

We may consider the clock to travel along the mutual x-axes of O and O' as in Figure 2 where, for our choice of axes, the sender is O and the receiver is O'. Now suppose a clock moves uniformly directly *towards* observer O in Figure 16.

Let us consider what the data taker sees as he looks at the approaching clock. He receives knowledge of the positions of the hands of the clock by the light emitted by the clock. Suppose, for the sake of simplicity, that the clock emits a light pulse towards the data taker with every tick. In the rest frame of the clock, two ticks will be separated by an interval of time T, say. In this rest frame of the clock the data taker's frame, O', will be approaching at velocity $-U$. The data taker, according to Equation 4, will

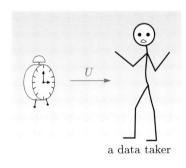

a data taker

Figure 16 Does the data taker see the clock as running fast or slow?

see two pulses that are separated by a time interval which he will measure to be

$$k(-U)\,T = T\,\sqrt{\dfrac{1 - \dfrac{U}{c}}{1 + \dfrac{U}{c}}}. \tag{41}$$

Objective 7   SAQ 27   What is the answer to the caption of Figure 16? Does this follow from Equation 41? How would the clock be seen to run if it were receding from the data taker?

We also discussed clocks from another standpoint in Section 3.2. According to the time dilation formula (Equation 26), 'moving clocks run slow'. If T is the interval between two ticks in the clock's rest frame and if the clock moves at velocity U in some inertial frame, then the interval between the same two ticks will be *observed* in that frame to be separated by an interval

$$\gamma(U)\,T = \dfrac{T}{\sqrt{1 - \left(\dfrac{U}{c}\right)^2}}. \tag{42}$$

Now one might be tempted to say that we have an inconsistency between these two experiments; for does not Equation 41 say that clocks run sometimes fast and sometimes slow, whereas Equation 42 says that moving clocks always run slow? Actually, of course, there can be no inconsistency here. In brief, the observations of the data taker in Figure 16 combine two effects: one is that which we have called time dilation and the other is the consequence of the fact that the light from the clock travels a shorter (or longer) distance every 'tick'. Once more, we are forced to think more carefully about the nature of the observations! Indeed, in Unit 5, we presented a way of establishing time throughout a frame — simply measuring the time of signals received from a body, particularly a moving body, will not do.

So, let us see how an observer could deduce from Equation 41 the fact that 'moving clocks run slow' according to Equation 42. In extracting information in this way, *the data taker can now represent the entire observing system,* for he is correcting for his particular position in spacetime. Briefly, he has to recognize that the time interval between the arrival of successive pulses is partly governed by the rate of ticking of the moving clock and partly by the fact that the two ticks were not emitted at the same position, so that the light pulses had different distances to travel. To unscramble the two he performs a calculation which allows for the latter effect.

In order to represent the observing system O′, the data taker must correct for the fact that he is visually observing events distant from him. He has established Equation 41 experimentally and he knows the clock moves at velocity U towards him. He considers two ticks of the clock. In his frame the first tick actually *occurs* at time t_1' and the next tick at t_2'. In the rest frame of the clock these ticks occur at t_1 and t_2. The data taker knows that the clock moves $U(t'_2 - t'_1)$ *closer* to him between the two ticks (Figure 17).

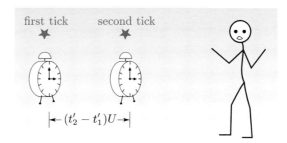

first tick   second tick

$|\!\leftarrow (t'_2 - t'_1)U \rightarrow\!|$

Figure 17   The clock moves closer between two ticks.

If the clock were motionless, these two pulses of light to his eye would need to cover the same distance and they would thus be received by him separated by the interval $(t'_2 - t'_1)$. But the second pulse is emitted a distance $U(t'_2 - t'_1)$ closer to him and thus arrives at his eye *sooner* by an amount $U(t'_2 - t'_1)/c$ than it would have done if the clock were stationary. Thus he reckons that the time interval separating the reception of the two pulses by his eye is

$$(t'_2 - t'_1) - \frac{U}{c}(t'_2 - t'_1) = \left(1 - \frac{U}{c}\right)(t'_2 - t'_1).$$

If $(t_2 - t_1)$ is the interval between ticks in the rest frame of the clock, then he knows from Equation 41 that the time interval between reception is also given by

$$(t_2 - t_1)\sqrt{\frac{1 - \dfrac{U}{c}}{1 + \dfrac{U}{c}}}.$$

Equating the two expressions gives

$$(t_2 - t_1)\sqrt{\frac{1 - \dfrac{U}{c}}{1 + \dfrac{U}{c}}} = (t'_2 - t'_1)\left(1 - \frac{U}{c}\right)$$

which simplifies to

$$(t'_2 - t'_1) = \frac{(t_2 - t_1)}{\sqrt{1 - \left(\dfrac{U}{c}\right)^2}}.$$

This is the usual time dilation formula of Equation 26. *Thus by taking account of the special character of his methods, he has arrived at an observation characteristic of his observing frame of reference, O'.*

Objective 7   SAQ 28   Follow through the above analysis for the case when the clock is receding from him.

Objective 7   SAQ 29   It has been said that 'moving metre sticks shrink'. Considering the observer to be an array of data takers situated at each spatial point in any frame, each with his synchronized clock, can you say exactly what Equation 24 in Section 3.1 describes?

4.2   The famous twin 'paradox'

Consider the following simplest possible statement of the so-called twin paradox. We have two identical twins at the origin of an inertial frame O. At time $t = 0$ one twin (called 'Jack the Nimble') takes off in a spaceship along the positive x-direction, where for simplicity we suppose that he accelerates impulsively (i.e. virtually instantly) to some constant velocity $\mathbf{V}$, say. The other twin (called John) stays throughout the entire process at his origin, $x = 0$. We may think of the ageing process in Jack the Nimble and in John as 'biological clocks'. Since these ageing processes depend, in the final analysis, on atomic processes, there is no fundamental difference between a 'biological clock' and an 'atomic clock' such as those used nowadays for time standards. For precision, therefore, you may want to consider that each has an identical clock. Now suppose that at some later time ($t = T/2$, say, on John's clock), Jack impulsively decelerates to velocity $-\mathbf{V}$ and comes back towards John. On John's clock, Jack will arrive home (after another sudden deceleration at $x = 0$) at time $t = T$. Jack's world-line as seen by John (or more precisely, perhaps, as seen by observing system O) is depicted in Figure 18.

To raise the 'paradox' into view let us calculate the time elapsed on Jack's clock between leaving and returning to John. In the first place you may question our use of special relativity, which only refers to inertial (i.e. non-accelerating) frames. But we may think of Jack the Nimble hopping onto an *inertial* system which is moving out along x at velocity $\mathbf{V}$ and subsequently hopping onto another inertial frame heading back with velocity $-\mathbf{V}$.

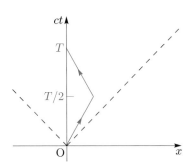

Figure 18   Jack the Nimble's world-line as observed by John in frame O.

The calculation is simple. The relevant equation is 26. According to John (observer O) the time out and back is T. Jack is observed by John, however, to be moving at speed V (velocity $+\mathbf{V}$ out and velocity $-\mathbf{V}$ back). Since John is an *inertial* observer throughout the whole process, he may apply the usual results of special relativity. John, therefore, knows that the time elapsed as shown on his clock will be reduced by a factor

$$\frac{1}{\gamma(V)} = \sqrt{1 - \left(\frac{V}{c}\right)^2}$$

on nimble Jack's clock. That is,

$$\begin{pmatrix} \text{time out and back} \\ \text{according to John} \end{pmatrix} = \frac{1}{\sqrt{1 - \left(\frac{V}{c}\right)^2}} \times \begin{pmatrix} \text{time out and back} \\ \text{according to Jack} \end{pmatrix}. \quad (43)$$

Thus, our prediction is that John's biological clock will have advanced further than Jack's, or in other words that John will have aged more than Jack.

The paradoxical nature of this analysis is the apparent symmetry in the problem. For could we not equally regard Jack as stationary and John as accelerating out and back so that now Jack ought to have aged more than John? The answer is, of course, no, for Jack and John are *not* equivalent observers: Jack has been the activist and has destroyed the symmetry by accelerating; he has had to change inertial frames three times; when he hops from one inertial frame to another he experiences an acceleration and so is distinguishable from John.

These effects are real and John *will* age more than Jack. The effect was verified experimentally in 1971 by Hafele and Keating who took very precise atomic clocks around the world in commercial airliners in order to compare their elapsed time after a flight with the elapsed time on a twin clock on the ground. In order to separate the velocity effect from the gravitational effect (to be discussed in Unit 9) two clocks were taken around the world, one in an an easterly and one in a westerly direction. The effect was confirmed to an accuracy of some 10%, but has been verified more precisely since.

In Unit 7, we shall analyse the motion of bodies undergoing *arbitrary* accelerations by using instantaneous rest frames. In particular we shall show that even if Jack undergoes an *arbitrary* programme of acceleration out and back he will still have aged less than John, though not by the same amount as is given by Equation 43 since this applies only to the case of Figure 18.

Objectives 4 and 8   SAQ 30   A realistic value for Jack's speed might be $10\,\mathrm{km\,s^{-1}}$. If John's elapsed time is 1 year, estimate nimble Jack's elapsed time. Take one year to be 3×10^7 seconds. Taylor's expansion,

$$\sqrt{1 - x} = 1 - \frac{x}{2} + (\text{terms of order } x^2 \text{ and higher})$$

is often useful when x is small compared with ± 1, and you may find it helpful here.

How much younger is Jack than John, by the time Jack returns?

---

5 Unit Summary

1 The Lorentz transformation formulae (Equations 23) refer to the situation in Figure 1 and relate the coordinates (t, x^1, x^2, x^3) and (t', x'^1, x'^2, x'^3) assigned to any event $\mathscr{E}$ by observers O and O' of Figure 1:

$$t' = \gamma(V)\left(t - \frac{V}{c^2}x^1\right), \tag{23a}$$

$$x'^1 = \gamma(V)(x^1 - Vt), \tag{23b}$$

$$x'^2 = x^2, \tag{23c}$$

$$x'^3 = x^3, \tag{23d}$$

$$\gamma(V) = \frac{1}{\sqrt{1 - \left(\dfrac{V}{c}\right)^2}}. \tag{23e}$$

Although we do not expect you to be able to reproduce the derivation of these equations, we do request that they and their meanings be committed to memory. One essential feature of the equations is that they 'mix up' the position and time of an event. Time in one inertial frame is a linear combination of time and distance in another; time is no longer separate from space. We now think in terms of 'spacetime'.

2 Midway in our derivation of the transformation formulae we derived an expression for the one-dimensional relativistic Doppler shift. According to Equation 5, electromagnetic waves emitted from one inertial frame and received in another are received at a higher frequency if the two frames approach each other along the line of transmission and at a lower frequency if the frames recede from each other. In Section 4.1 it is also pointed out that because of the Doppler effect, a moving clock when viewed by eye or photographed, appears to run fast as it approaches and slow as it recedes.

3 Two applications of the transformation equations are the famous length contraction and time dilation formulae, Equations 24 and 26 respectively:

$$l = l_0\sqrt{1 - \left(\frac{V}{c}\right)^2} = \frac{l_0}{\gamma(V)}, \tag{24}$$

$$\tau = \tau_0\frac{1}{\sqrt{1 - \left(\dfrac{V}{c}\right)^2}} = \gamma(V)\,\tau_0. \tag{26}$$

You ought to be able to derive these equations and describe the measurement process to which they refer. Concerning time, it was shown in Section 4.1 how visual observation can be disentangled from the relativistic Doppler effect to yield the usual time dilation equation.

4 A somewhat more involved application of the transformation formulae was given in Section 3.3 where we showed that if a particle is moving as in Figure 12, with velocity U^1 in the x^1-direction according to O and velocity U'^1 in the x'^1-direction according to O', then U^1 and U'^1 are connected by Equation 29:

$$U'^1 = \frac{U^1 - V}{1 - \dfrac{VU^1}{c^2}}. \tag{29}$$

More generally, the particle may also have velocity components U^2 and U^3 according to O and U'^2 and U'^3 according to O′, and these are related by Equations 31 and 32:

$$U'^2 = \frac{U^2}{\gamma(V)\left(1 - \dfrac{U^1 V}{c^2}\right)}, \qquad (31)$$

$$U'^3 = \frac{U^3}{\gamma(V)\left(1 - \dfrac{U^1 V}{c^2}\right)}. \qquad (32)$$

5 It was shown in Section 3.3 that it is impossible to select the velocities of any sequence of inertial frames, each moving at a speed $< c$ with respect to its preceding member of the sequence, in such a way that the last frame moves faster than c with respect to the first frame.

6 It follows from the time dilation formula (Equation 26, see item 3 in this Summary) that fast-moving unstable fundamental particles (e.g. the muons of SAQ 17) move farther between their creation and decay than they would be able to if the Newtonian world-view applied.

7 In Section 4.2, application of Equation 26 (see item 3 in this Summary) to the twin paradox led to the unavoidable conclusion that a twin who leaves his brother and returns, has aged less. This story will be continued in more general terms in Unit 7.

8 In Section 3.4 we showed that the assumption of charge invariance in our thought experiment of Section 2 of Unit 5 was consistent with the Einstein world-view. That both length and time are relative to the observer is underlined by this thought-experiment (see SAQ 25), and you are expected to be able to describe it (see Objectives).

Band 6 of AC2 comments on this Unit.

Self-assessment questions — answers and comments

SAQ 1 For two inertial frames, as in Figure 2, the Galilean transformation is $t' = t$ (time being absolute) and $x' = x - Vt$, where V is the velocity of O' relative to O.

SAQ 2 The Galilean transformation implies that the velocity of anything as seen in one inertial frame is related to the velocity in a uniformly moving frame by adding the relative velocity of the two frames. But this consequence of the Galilean transformation is in contradiction with SR2 which says that light has speed c in all inertial frames.

SAQ 3 The world-line is shown in Figure 19. Since O' moves to the right, with velocity **V**, as observed by O, the situation would be reversed if O' were observing O.

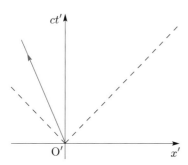

Figure 19   O moves to the left relative to O'.

SAQ 4 This is a review of material in Unit 5. There, in SAQ 23, you saw that one could assign a time to a distant event by supposing it to be triggered by a pulse sent out at time t_{out} and returned from the distant event to be received back at time t_{back}. In Unit 5 we had for the time assigned to the event

$$t_{\text{event}} = \tfrac{1}{2}\left(t_{\text{out}} + t_{\text{back}}\right).$$

One can also write the position of the event in terms of t_{out} and t_{back} by realizing that c is the same for all directions. The time of travel for the light pulse, from transmission to reception, is $(t_{\text{back}} - t_{\text{out}})$. Thus the total distance covered is $c(t_{\text{back}} - t_{\text{out}})$. The distance to the 'turn around' point, which locates the event, is $\tfrac{1}{2}c(t_{\text{back}} - t_{\text{out}})$. Referring to Figure 6 and writing t_1 for t_{out} and t_3 for t_{back} gives the spacetime coordinates of $\mathscr{E}$ as

$$(ct, x) = \left(\frac{c(t_1 + t_3)}{2}, \frac{c(t_3 - t_1)}{2}\right).$$

SAQ 5 From Equation 4,

$$k(V) = \sqrt{\frac{1 + V/c}{1 - V/c}}.$$

Clearly, $k(0) = 1$.

If V/c is not negative then $(1 + V/c)$ is equal to or greater than 1, but $(1 - V/c)$ is equal to or less than 1. Thus the quantity inside the square root is not less than 1. But the square root of a number exceeding 1 also exceeds 1. Thus if $0 \leqslant V/c < 1$, then $k(V) \geqslant 1$. Notice we must exclude $V/c = 1$ since $k(V)$ is not defined for this value of V (it would be the square root of 2 divided by zero!).

Similarly, if V/c is negative and lies between 0 and -1 we can say that $(1 + V/c)$ is less than 1 but greater than 0 while $(1 - V/c)$ is greater than 1. Thus the quantity inside the square root is now less than 1 but greater than 0. The square root of such a number is always less than 1. Thus, if $-1 < V/c < 0$ we have $0 < k(V) < 1$.

SAQ 6 Figure 20 is a graph showing how k depends on V/c. If O and O' are moving away from each other, then $k > 1$. If they are approaching one another, then $k < 1$.

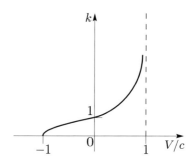

Figure 20   The factor k tends to ∞ as $V/c \longrightarrow 1$, indicated by the dotted line for $V/c = 1$ to which the curve is asymptotic.

SAQ 7 According to Equation 5,

$$f' = f\sqrt{\frac{1 - V/c}{1 + V/c}}.$$

In this equation, V is the velocity of O' as observed by O. Thus a positive V means that O' moves away from O, i.e. they recede from each other. An analysis similar to that of SAQ 5 then shows that the quantity within the square root is less than 1, so $f' < f$.

On the other hand if V is negative then O' moves towards O. For negative values of V the quantity inside the square root is greater than 1 and $f' > f$.

When $V = 0$, of course, there is no shift and $f = f'$.

SAQ 8 The graph should be as shown in Figure 21.

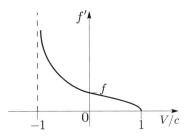

Figure 21 The relativistic Doppler shift; the frequency may be raised ($f' > f$) or lowered ($f' < f$). As $V \longrightarrow -c$, $f' \longrightarrow \infty$, as indicated by dotted line to which the curve is asymptotic.

SAQ 9 This is a matter of algebraic manipulation. The expression for $k(V)$ is given in Equation 4:

$$k(V) = \sqrt{\frac{1 + V/c}{1 - V/c}}. \qquad (a)$$

The two equations relating t and x to t' and x' are: Equation 13, which we slightly reorder to read,

$$ct' + x' = \frac{1}{k(V)}(ct + x) \qquad (b)$$

and Equation 17, which reads

$$ct' - x' = k(V)(ct - x). \qquad (c)$$

First add Equations b and c and divide by $2c$:

$$t' = \frac{1}{2c}\left\{ \frac{1}{k(V)}(ct + x) + k(V)(ct - x) \right\}.$$

This can be written

$$t' = \frac{1}{2}\left(k(V) + \frac{1}{k(V)} \right) t - \frac{1}{2c}\left(k(V) - \frac{1}{k(V)} \right) x. \quad (d)$$

Next subtract Equation c from Equation b and divide by 2:

$$x' = \frac{1}{2}\left(k(V) + \frac{1}{k(V)} \right) x - \frac{c}{2}\left(k(V) - \frac{1}{k(V)} \right) t. \quad (e)$$

It remains merely to express $k(V)$ in terms of V/c using Equation a. We have, writing k for $k(V)$:

$$k + \frac{1}{k} = \sqrt{\frac{1 - V/c}{1 + V/c}} + \sqrt{\frac{1 + V/c}{1 - V/c}}$$

$$= \frac{(1 - V/c) + (1 + V/c)}{\sqrt{(1 - V/c)(1 + V/c)}}$$

$$= \frac{2}{\sqrt{1 - V^2/c^2}} = 2\gamma.$$

Similarly,

$$k - \frac{1}{k} = 2\frac{V}{c}\gamma.$$

Hence, from Equation d and Equation e we get the customary form of the Lorentz transformation:

$$t' = \gamma(V)\left(t - \frac{V}{c^2}x \right)$$

$$x' = \gamma(V)(x - Vt)$$

where

$$\gamma(V) = \frac{1}{\sqrt{1 - \left(\dfrac{V}{c}\right)^2}}.$$

SAQ 10 When V/c is negligibly small, $\gamma(V)$ is essentially 1.

In Equation 19, V/c only occurs in the term for $\gamma(V)$, so in the non-relativistic limit $x' = x - Vt$.

This is the Galilean form.

SAQ 11 Equation 18 has V/c occurring in two places: in $\gamma(V)$ which becomes 1 when V/c is small, and in $(V/c^2)x$ which becomes zero when V/c is small and x is finite. Thus, when V/c is negligibly small, t' and t are essentially equal — we recover Newton's absolute time.

SAQ 12 For the five given values of $|V|$, $\gamma(V)$ is respectively: 1.00005, 1.005, 1.048, 3.203, 10.01. If $|V|$ can never exceed c, $-1 < V/c < +1$. A graph of $\gamma(V)$ versus V/c would therefore be as shown in Figure 22.

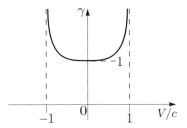

Figure 22 How γ depends on V/c.

SAQ 13 We wish to solve Equations 18 to 20 for t and x in terms of t' and x'. From Equations 18 and 19 we have:

$$x' + Vt' = \gamma(V)(x - Vt) + \gamma(V)\,V\left(t - \frac{V}{c^2}x \right)$$

$$= \left(1 - \frac{V^2}{c^2} \right)\gamma(V)\,x.$$

But from Equation 20:

$$\gamma(V) = \frac{1}{\sqrt{1 - \left(\dfrac{V}{c}\right)^2}}.$$

Thus

$$x' + Vt' = \frac{1}{\gamma(V)}x$$

or

$$x = \gamma(V)(x' + Vt').$$

Similarly, again from Equations 18 and 19,

$$t' + \frac{V}{c^2}x' = \gamma(V)\left(t - \frac{V}{c^2}x\right) + \frac{V}{c^2}\gamma(V)(x - Vt)$$

$$= \gamma(V)\left(1 - \frac{V^2}{c^2}\right)t = \frac{1}{\gamma(V)}t$$

or $\qquad t = \gamma(V)\left(t' + \frac{V}{c^2}x'\right).$

SAQ 14 **(a)** Equations 23 relate the spacetime coordinates of a single event as recorded by two inertial observers O and O' whose relative uniform motion lies along their mutual collinear x- and x'-axes. O' moves with velocity V relative to O. Equations 23 apply to the case that the origins of O and O' coincide at time $t = t' = 0$.

(b) $x^1 = \gamma(V)(x'^1 + Vt')$
$\qquad x^2 = x'^2$
$\qquad x^3 = x'^3$
$\qquad t = \gamma(V)\left(t' + \frac{V}{c^2}x'^1\right).$

(c) $ct' = \gamma(V)\left(ct - \frac{V}{c}x^1\right)$

$\qquad x'^1 = \gamma(V)\left(x^1 - \frac{V}{c}ct\right).$

You might feel that the two equations look more like each other when written in this form.

SAQ 15 **(a)** Equation 24 indicates that if the positions of the two ends of a rod are observed from a frame moving relative to the rod, and the two observations are made *at the same time* in this frame, then the length of the rod as deduced from these observations is reduced by a factor $\sqrt{1 - (V/c)^2}$ compared with the rest length l_0.

(b) When V/c is negligibly small then $l = l_0$ to a good approximation. If, on the other hand, $(V/c)^2$ is nearly unity then the rod will appear to be compressed towards zero length when observed from the moving frame described in (a) above.

SAQ 16 According to Equations 23c and 23d, coordinates *perpendicular* to the motion are unaffected. The observed length of a rod moving perpendicular to its length is therefore the same as its rest length.

SAQ 17 In its rest frame, a muon decays in time 2×10^{-6} s. If its speed is $0.95c$, its observed decay time (see Equation 26) is:

$$\tau = 2 \times 10^{-6} \times \frac{1}{\sqrt{1 - (0.95)^2}} = 6.4 \times 10^{-6}\,\text{s}.$$

It will be observed to travel a distance

$$l = (0.95)(3 \times 10^8)(6.4 \times 10^{-6}) = 18.2 \times 10^2\,\text{m}$$

before decaying. This is more than three times the distance travelled according to Newtonian physics.

SAQ 18 The inverse equation is Equation 22. The primed coordinates correspond to the rest frame of the clock, so that two successive ticks correspond to the same value of x', i.e. $x'_B - x'_A = 0$, in the notation of Section 3.2. Hence:

$$t_B - t_A = \gamma(V)[(t'_B - t'_A) - (V/c)0]$$

which leads directly to Equation 26.

SAQ 19 **(a)** Two inertial observers have their x^1- and x'^1-axes aligned, with O' moving with velocity V relative to O (V may be positive or negative). If a particle moves such that its x^1-component of velocity as observed by O is U^1, then U'^1, the particle's x'^1-component of velocity as observed by O', is given by Equation 29.

(b) In the non-relativistic limit (U^1V/c^2 small) the equation is effectively just $U'^1 = U^1 - V$. This is the usual Galilean equation; in everyday terms it simply says that if we are moving slowly along the path of a particle, then the particle appears to us to be moving faster or slower according to the Newtonian expression for calculating relative velocities.

(c) Inertial frames are equivalent; the only distinction for velocity transformation can be in the sign of V. We have then

$$U^1 = \frac{U'^1 + V}{1 + \dfrac{V}{c^2}U'^1}.$$

Of course, if you solve Equation 29 for U^1 in terms of U'^1 you will get this result. To be specific, suppose we have Equation 29:

$$U'^1 = \frac{U^1 - V}{1 - \dfrac{V}{c^2}U^1}.$$

Then we may multiply both sides by $\left(1 - \dfrac{V}{c^2}U^1\right)$ to obtain

$$U'^1 - \frac{V}{c^2}U^1U'^1 = U^1 - V.$$

Collecting the terms involving U^1 gives:

$$U^1\left(1 + \frac{V}{c^2}U'^1\right) = U'^1 + V$$

or

$$U^1 = \frac{U'^1 + V}{1 + \dfrac{V}{c^2}U'^1}.$$

SAQ 20 Using Equations 23a and 23c, the U'^2-component (for O') is:

$$U'^2 = \frac{\Delta x'^2}{\Delta t'} = \frac{\Delta x^2}{\gamma(V)\left(\Delta t - \dfrac{V}{c^2}\Delta x^1\right)}.$$

Dividing top and bottom by Δt gives:

$$U'^2 = \frac{\dfrac{\Delta x^2}{\Delta t}}{\gamma(V)\left(1 - \dfrac{V}{c^2}\dfrac{\Delta x^1}{\Delta t}\right)} = \frac{U^2}{\gamma(V)\left(1 - \dfrac{U^1 V}{c^2}\right)}.$$

Similarly for the x^3- and x'^3-directions, Equations 23a and 23d give

$$U'^3 = \frac{U^3}{\gamma(V)\left(1 - \dfrac{U^1 V}{c^2}\right)}.$$

SAQ 21 One readily finds that $U^1 = c$. You cannot catch up with light; however fast you go, it still has speed c.

SAQ 22 This is an application of Equation 29. Here, observer O$'$ moves with velocity u relative to O. Thus, in Equation 29, we must take $V = u$. Furthermore, in frame O, the electrons move with x-component of velocity $-u$. Thus, in Equation 29, we must take $U^1 = -u$. The result is (with U'^1 now denoted by v),

$$v = \frac{-u - u}{1 - \dfrac{u}{c^2}(-u)} = \frac{-2u}{1 + \left(\dfrac{u}{c}\right)^2}.$$

SAQ 23 That $\lambda_a'^{+} > \lambda_0$ follows from the fact that $(u/c)^2 > 0$. Thus

$$\sqrt{1 - \left(\frac{u}{c}\right)^2} < 1.$$

SAQ 24 From Equation 38 we have:

$$\lambda_b'^{-} = \frac{\lambda_a'^{-}}{\sqrt{1 - \left(\dfrac{v}{c}\right)^2}}.$$

But $\lambda_a'^{-}$, is given by Equation 36 and v is given by Equation 34. Thus

$$\lambda_b'^{-} = \frac{\lambda_0 \sqrt{1 - \left(\dfrac{u}{c}\right)^2}}{\sqrt{1 - 4\dfrac{\left(\dfrac{u}{c}\right)^2}{\left(1 + \dfrac{u^2}{c^2}\right)^2}}}.$$

The term in the square root of the denominator is

$$\frac{\left(1 + \dfrac{u^2}{c^2}\right)^2 - 4\dfrac{u^2}{c^2}}{\left(1 + \dfrac{u^2}{c^2}\right)^2} = \frac{1 - 2\dfrac{u^2}{c^2} + \dfrac{u^4}{c^4}}{\left(1 + \dfrac{u^2}{c^2}\right)^2} = \frac{\left(1 - \dfrac{u^2}{c^2}\right)^2}{\left(1 + \dfrac{u^2}{c^2}\right)^2}$$

giving

$$\lambda_b'^{-} = \lambda_0 \frac{\sqrt{1 - \left(\dfrac{u}{c}\right)^2}}{\sqrt{\dfrac{\left(1 - \dfrac{u^2}{c^2}\right)^2}{\left(1 + \dfrac{u^2}{c^2}\right)^2}}} = \lambda_0 \frac{\left(1 + \dfrac{u^2}{c^2}\right)}{\sqrt{1 - \dfrac{u^2}{c^2}}}.$$

SAQ 25 The mobile electrons are spread out over the whole length of the circuit. Two different inertial observers (who move with respect to each other) will disagree on the simultaneity of events that are spatially separated. For instance, consider the following two events, $\mathscr{E}_a$ and $\mathscr{E}_b$:

$\mathscr{E}_a$ is the event that an electron in wire 'a' leaves wire 'a' for wire 'b' (see Figure 15);

$\mathscr{E}_b$ is the event that some other electron in wire 'b' leaves wire 'b' for wire 'a'.

Clearly, the relative times of these two events will be observed differently by O and by O$'$. It is not too surprising, then, that if O and O$'$ cannot agree on when the electrons change over from being on one wire to being on the other, they are not going to agree on the number of electrons in the two wires.

The positive charges, on the other hand, are *fixed* to their respective wires. Due to the apparent length-contraction effects of special relativity, O and O$'$ will disagree on where the individual positive charges are, but they will agree on the total number in each wire.

SAQ 26 We set out in Section 3.4 to show that our assumption of charge invariance (setting $q' = q$ in Equations 2a and 2b of Unit 5) was consistent with the subsequent argument in Section 2 of Unit 5 that led to the conclusion that length is relative to the observer. By showing that the total number of electrons in the circuit is the same for inertial observers O and O$'$, we have succeeded. In the process (SAQ 25) we saw that the explanation lies in the relativity of simultaneity.

SAQ 27 If the clock approaches the data taker then $U > 0$. The interval between two ticks of the clock is therefore reduced, according to Equation 41, since $(1 - U/c) < 1$ and $(1 + U/c) > 1$; in other words, the clock of Figure 16 is *seen* by the data taker to run fast. If the clock were receding then U would be negative so $(1 - U/c) > 1$ and $(1 + U/c) < 1$ and the clock would be *seen* to run slow. In both cases, however, the *observed* time would run slow according to the time dilation formula (Equation 26); U^2 is positive, whether U itself is positive or negative. (See also the continuing discussion in the text.)

SAQ 28 The analysis preceding this question makes no distinction between positive and negative values of U. The only changes needed are that 'closer to' becomes 'farther from' and 'sooner' becomes 'later'. The result is unchanged (again, U^2 is positive whether U itself is positive or negative).

SAQ 29 The array of data takers have identical synchronized clocks. Each relevant member of this group records which part of the metre stick is at his location at a pre-assigned time as read on his clock.

Collecting results afterwards shows that, at that pre-assigned time, the rod was observed to be shortened by the factor $1/\gamma(V)$, where V is the rod's velocity relative to the data takers.

It is interesting that if a single data taker actually views the moving rod, with his *eyes*, he will *see* the rod appear to be rotated spatially rather than simply shrunk.

SAQ 30 Jack's elapsed time

$$= \sqrt{1 - \left(\frac{V}{c}\right)^2} \times 1 \text{ year}$$

$$= \sqrt{1 - \left(\frac{10}{3 \times 10^5}\right)^2} \text{ year}$$

$$\approx 1 - \tfrac{1}{2}(\tfrac{1}{9})10^{-8} \text{ year}.$$

Jack is younger than John by about

$$\frac{1}{18}(10^{-8}) \text{ year} \approx \frac{10^{-8}}{18} \times 3 \times 10^7 \,\text{s}$$

$$\approx \frac{1}{60} \,\text{s}.$$

This should *not* be construed to mean that relativistic effects are always small. In fundamental particle physics they are normally dominant.

Unit 7 Spacetime, momentum and energy

Prepared by the Course Team

Contents

Aims

In this Unit we intend to:

1 Show that the fundamental properties of space and time lead to the conclusion that the world is four-dimensional, and that this in turn leads to the introduction of an invariant interval (analogous to distance in three-dimensional space) and necessitates a re-examination of the concepts of past, present and future.

2 Define the special relativistic forms of momentum and energy.

Objectives

When you have finished studying this Unit, you should be able to:

1 State Theorems I–IV in your own terms, explain their meaning, and use them to solve simple problems.

2 Write down the invariant interval in two-dimensional spacetime and describe its significance; using it, give an argument showing that for two events (an event pair), one event may lie in the *absolute future*, *absolute past*, or *elsewhere* with respect to the other, explaining these terms and the terms *time-like* and *space-like* in the process.

3 Describe the world-lines (on spacetime diagrams) of particles in two-dimensional spacetime; define the term *instantaneous rest frame*, and use it to explain what is meant by the *proper time* of a particle moving with arbitrary velocity (less than c) and arbitrary acceleration.

4 Describe the characteristic features of the relativistic Doppler shift and time dilation effect for a particle moving in three-dimensional space, solving simple problems if given the relevant equations.

5 Give the special relativistic expressions for the linear momentum and the energy of a free particle. Use the 'superscript zero' notation.

6 Write the Lorentz transformation connecting the pair (x^0, x^1) with (x'^0, x'^1) and the pair (p^0, p^1) with (p'^0, p'^1) for any two inertial observers O and O'. Write the other details of Figure 24 (especially the Lorentz invariance of $(p^0)^2 - (p^1)^2$), explaining the meaning of the equations and using them to solve simple problems.

7 Explain the physical and theoretical significance of the space component p^1, and time component p^0, of the momentum of a particle, especially in the context of collisions between particles (see points (i)–(iv) of Section 6.5).

8 Perform calculations involving transformations between matter and energy using $E = mc^2$ or otherwise.

9 Give an account of the so-called twin paradox and its relation to the Theorems I–IV.

1 Introduction to Unit 7

Band 7 of AC2 introduces this Unit.

In this Unit we concentrate on two main topics. First, we dwell on the manner in which time and space are inextricably commingled; the world is four-dimensional and events need to be labelled by *four* variables (x^0, x^1, x^2, x^3), and we consider how to regard the motion of particles in spacetime. The motion of particles leads naturally to the second main topic: the relativistic modification of Newtonian linear momentum. We consider how this leads to a modification also of the Newtonian expression for the energy of a free particle.

Along the way, we introduce a number of concepts which will be of great importance when we study general relativity in the following Block. At the end of the Unit, we shall draw attention to these important concepts.

As an example of the dramatic consequences of special relativity, we briefly touch on the famous expression $E = mc^2$.

Study comment

What to do if you get behind
This is the longest Unit of the block, and the Course Team recognize the need for a strategy if you get behind in your study.

In the first place, you will find a longish passage in Section 3.2 which is completely optional. In this passage we include a proof of Theorem I for completeness, but no assessment will be based on it.

In addition, there are two passages in Section 6.5, both fairly demanding, which are marked as 'optional reading' but in this case the meaning of 'optional' is not quite so clear cut. These passages can be omitted without prejudicing in any way your study of later Units. However, you will notice that three SAQs fall within the second of these passages. If you omit these passages, you may not meet the related objectives in full, but the effect on your ability to do any assessment will be small.

2 Preliminaries

Several far reaching concepts have been introduced in the previous two Units.

First, the concept of an *event* has been pivotal to our discussion, especially in Unit 6. There we emphasized that observers viewing the same event from different inertial frames will assign different values to the coordinates.

Second, by applying several *additional* basic assumptions to the two postulates SR1 and SR2 in Unit 5, we arrived at the Lorentz transformation equations connecting the coordinates of an event $\mathscr{E}$ according to two inertial observers, O and O', whose x- and x'-axes are collinear, as in Figure 1. (Recall that where no confusion can arise, and x^2, x^3 are not being considered, we write x for x^1.)

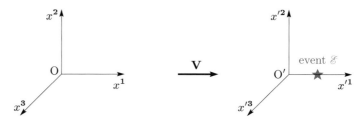

Figure 1   Two inertial frames moving uniformly relative to one another along their collinear **1**-axes.

Since we shall be needing these transformation equations throughout the Unit, we reproduce them here. For events occurring on the x- (and x'-) axis only, the Lorentz transformation equations are:

$$x' = \gamma(V)(x - Vt) \tag{1a}$$

$$t' = \gamma(V)\left(t - \frac{V}{c^2}x\right) \tag{1b}$$

where

$$\gamma(V) = \frac{1}{\sqrt{1 - \left(\dfrac{V}{c}\right)^2}}. \tag{1c}$$

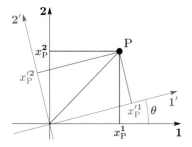

Figure 2   A rotation about the **3**-direction by an angle θ.

Equations 1a and 1b emphasize the way that x and t get 'mixed up' together in special relativity.

The Lorentz transformation equations are *linear* equations connecting (t, x) to (t', x'); that is, they contain no terms like x^2 or xt or t^2, or higher powers of x and t. You have already met this great simplifying feature in another context, that of rotations of, say, the x^1- and x^2-axes about the x^3-direction in our well-accustomed Euclidean space of everyday experience. Figure 2 shows how the coordinates (x_P^1, x_P^2) and $(x_P'^1, x_P'^2)$ of some stationary point P are related to each other when the x^1- and x^2-axes are rotated by an angle θ about the x^3-axis, to form new axes x'^1 and x'^2.

In Unit 1 you saw that the equations connecting such coordinates are linear equations. You also know well, and it is clear from Figure 3, that any such rotation of axes in Euclidean space leaves unchanged the *distance* d_{AB} between any two points P_A and P_B.

We thus have the parallel that the Lorentz transformation Equations 1a and 1b are *linear* equations connecting the coordinates (t, x) and (t', x') of

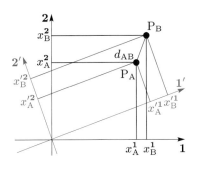

Figure 3   The distance between any two points is invariant with respect to rotation of axes.

an event $\mathscr{E}$, just as the coordinates of a point P for rotations of axes in Euclidean space, (x^1, x^2) and (x'^1, x'^2), are also connected by *linear* equations.

In the latter case there is a well-defined distance d_{AB} between any two points P_A and P_B in the (x^1, x^2) plane, for by Pythagoras' theorem:

$$d_{AB} = \sqrt{(x_A^1 - x_B^1)^2 + (x_A^2 - x_B^2)^2}. \tag{2a}$$

Similarly, the distance between points P_A and P_B in the (x'^1, x'^2) plane is given by:

$$d'_{AB} = \sqrt{(x_A'^1 - x_B'^1)^2 + (x_A'^2 - x_B'^2)^2}. \tag{2b}$$

Since this distance is unaffected by rotations of axes x^1 and x^2 about x^3, we can say:

$$d_{AB} = d'_{AB}$$

or

$$\sqrt{(x_A^1 - x_B^1)^2 + (x_A^2 - x_B^2)^2} = \sqrt{(x_A'^1 - x_B'^1)^2 + (x_A'^2 - x_B'^2)^2}.$$

What is the analogous concept for spacetime? If we can define a 'distance' between any two events $\mathscr{E}_a$ and $\mathscr{E}_b$ in spacetime we shall have a structure analogous to the usual Euclidean space. Such a definition is possible, giving rise to the mathematical structure called Minkowski spacetime.

3 The invariant interval

3.1 Two-dimensional spacetime

spacetime

We define *spacetime* as the set of all possible events; the points in spacetime are events. In the special theory of relativity, any event occurring on the x-axis (in space) may validly be assigned coordinates (ct, x) (in spacetime) by any *inertial* observer, and all *inertial* observers are equivalent by postulate SR1 (Unit 5). The Lorentz transformation Equations 1a, 1b, and 1c provide the necessary connection between any two observers such as those depicted in Figure 1.

Lorentz invariant

We now define a 'distance' between any two events in such a way that any such 'distance' is *Lorentz invariant*; that is, it is the *same* for all inertial observers. The justification for defining distance in the way we are about to do is simply that the new concept will be very *useful*. You will by now be used to the idea that concepts often prove useful because they are conserved under a transformation.

A clue to the most reasonable definition of such a 'distance' is provided by postulate SR2: The speed of light in free space has the same value in all inertial frames. In other words it is a Lorentz invariant quantity. To introduce a Lorentz invariant concept of distance, we first consider two events that are connected by a light pulse. Events $\mathscr{E}_a$ and $\mathscr{E}_b$ in Figure 4 are located on the x- (and x'-) axis of space (but since $t \neq 0$ this is not the x-axis of spacetime). The 'distance' between the events is to be expressed in terms of the spacetime coordinates (ct_a, x_a) and (ct_b, x_b) assigned by an inertial observer O.

Being connected by a light pulse, $\mathscr{E}_a$ and $\mathscr{E}_b$ in Figure 4 are a special pair of events; the same could not be said, for example, of the pairs $(\mathscr{E}_a, \mathscr{E}_d)$ or $(\mathscr{E}_b, \mathscr{E}_c)$. But since the pair $(\mathscr{E}_a, \mathscr{E}_b)$ are so connected, we can write

$$x_b - x_a = c(t_b - t_a).$$

Similarly, there are other events connected by a light pulse to event $\mathscr{E}_a$. For example, events $\mathscr{E}_a$ and $\mathscr{E}_c$ are connected by the equation

$$x_c - x_a = -c(t_c - t_a).$$

In general, we can say that all events $\mathscr{E}_A$ connected to event $\mathscr{E}_a$ by a light pulse will obey the equation

$$x_A - x_a = \pm c(t_A - t_a). \tag{3a}$$

Interpreting this result for any other inertial observer O' (whose x'-axis is collinear with the x-axis as in Figure 1) we have

$$x'_A - x'_a = \pm c(t'_A - t'_a). \tag{3b}$$

These two equations express the facts that $\mathscr{E}_A$ and $\mathscr{E}_a$ are connected by a light pulse and that c is a Lorentz invariant.

The 'distance' between $\mathscr{E}_A$ and $\mathscr{E}_a$ will most simply be expected to involve coordinate differences $(x_A - x_a)$ and $c(t_A - t_a)$, by analogy with Equations 2 for Euclidean space. Furthermore, again by analogy, we expect the distance will not change value if x_A and x_a, or t_A and t_a, are interchanged. The simplest assumption is that the 'distance' between $\mathscr{E}_A$ and $\mathscr{E}_a$ depends only on the squares $(x_A - x_a)^2$ and $(t_A - t_a)^2$, cf. Equation 2 in Euclidean space. But Equations 3a and 3b must be satisfied for events connected by a light pulse.

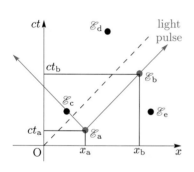

Figure 4 Some event pairs are connected by a light pulse while others are not.

From Equation 3a, the only relationship between the squares $(x_A - x_a)^2$ and $(t_A - t_a)^2$ is:

$$(x_A - x_a)^2 = c^2(t_A - t_a)^2.$$

Rewriting this equation gives:

$$(ct_A - ct_a)^2 - (x_A - x_a)^2 = 0.$$

Similarly, Equation 3b says that the following relationship holds for any other inertial observer (with collinear **1**-axis):

$$(ct'_A - ct'_a)^2 - (x'_A - x'_a)^2 = 0.$$

Thus far, we have only managed to show that if any pair of events *is connected by a light signal* then all inertial observers (whose x-axes are collinear) will agree that the difference between the squares of the time coordinates and of the x-coordinates of the two events is zero. But what we really seek is a Lorentz invariant definition of 'distance' between *any* pair of events in spacetime and not just between events connected by a light signal. In Figure 4, for example, how would we define 'distance' between events $\mathscr{E}_a$ and $\mathscr{E}_d$ or between the pair $\mathscr{E}_d$ and $\mathscr{E}_e$? We first propose a definition of 'distance' and then confirm that it is indeed a Lorentz invariant definition. We *define* the square of the 'distance' between *any* two events $\mathscr{E}_a$ and $\mathscr{E}_b$ as

$$(S_{ab})^2 = (ct_b - ct_a)^2 - (x_b - x_a)^2 \qquad (4)$$

where of course $(S_{ab})^2 = 0$ if $\mathscr{E}_a$ and $\mathscr{E}_b$ are connected by a light pulse. To emphasize that we are now in the domain of Minkowski spacetime and not Euclidean space, we shall henceforth refer to the 'distance' S_{ab} as the *invariant interval* between $\mathscr{E}_a$ and $\mathscr{E}_b$.

invariant interval

The 'distance' or 'interval' defined in Equation 4 has some properties which are not shared by what we call distance in everyday speech. For example, the interval S_{ab} between two well-separated events connected by a light pulse is zero, and the square $(S_{ab})^2$ can even be negative for certain pairs $\mathscr{E}_b$ and $\mathscr{E}_a$. However, as we now show, $(S_{ab})^2$ is indeed a Lorentz invariant for *any* event pair whose coordinates transform according to the Lorentz transformation, Equations 1a, 1b, and 1c. We consider two proofs of the Lorentz invariance of $(S_{ab})^2$.

The first proof involves a straightforward manipulation of the Lorentz transformation equations, and we ask you to try to deduce it in SAQ 1. This proof refers specifically to two events occurring on the x^1- and x'^1-axes for observers O and O$'$ moving relatively along their x^1- and x'^1-axes. The second proof is in some ways more elegant, in that these restrictions are lifted. However, we shall defer it to the end of the next Section since we must first set the scene by introducing general inertial frames in four-dimensional spacetime.

Objective 2   SAQ 1   Suppose two arbitrary events $\mathscr{E}_a$ and $\mathscr{E}_b$ occur on the collinear x- and x'-axes of the inertial observers O and O$'$ of Figure 1. Using the Lorentz Equations 1a and 1b, connecting coordinates (ct_a, x_a) with (ct'_a, x'_a) and (ct_b, x_b) with (ct'_b, x'_b), show that $(S_{ab})^2$ as defined by Equation 4 is a Lorentz invariant; that is, show that

$$(ct_b - ct_a)^2 - (x_b - x_a)^2 = (ct'_b - ct'_a)^2 - (x'_b - x'_a)^2.$$

3.2   Four-dimensional spacetime (Theorem I)

As you know, the specification of any event $\mathscr{E}$ by an inertial observer requires four coordinates ct, x^1, x^2 and x^3. For the case of Figure 5 you have seen that the

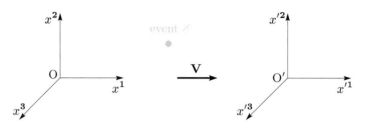

Figure 5   An event may occur anywhere in spacetime.

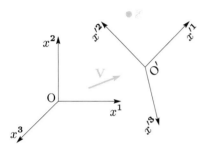

Figure 6   The axes of observer O′ are rotated (but not rotating) and displaced with respect to those of observer O, and O′ moves uniformly at an arbitrary velocity **V** relative to O.

coordinates assigned to $\mathscr{E}$ by O and by O′ are as in Equations 1a and 1b, with the additional equations

$$\begin{cases} x^2 = x'^2 \\ x^3 = x'^3 \end{cases} \quad . \tag{5}$$

But even though we are now allowing the event to occur off the **1**-axes, this is still not the most general case. As you know, in Newtonian physics the *only* restriction on pairs of inertial frames is that they must be in uniform relative translational motion or lie at rest with respect to one another. (Of course, it is also required that inertial frames do not accelerate or rotate.) The most general case considered in special relativity is that illustrated in Figure 6, where not only is event $\mathscr{E}$ off the **1**-axes, but the axes of O and O′ are rotated (but *not* rotating) relative to one another, and the relative translation of O and O′ occurs in an arbitrary direction.

Indeed, the full Lorentz transformation for the general case depicted in Figure 6 can be worked out and is similar in two ways to Equations 1a, 1b and 5, which apply to an event taking place anywhere off the x^1- and x'^1-axes in Figure 5.

In the first place, the connection between coordinates assigned to $\mathscr{E}$ by O and O′ of Figure 6 are *linear* equations, the difference is one of complexity with the four rather simple linear equations (1a), (1b) and (5), replaced by four less simple linear equations each in general involving all four spacetime coordinates.

In the second place, there is a generalization of the invariant interval, Equation 4 (which follows from Equations 1a, 1b and 5 — see SAQ 1), appropriate to the general case of Figure 6.

It is this second property of the general Lorentz transformation that we now consider. We seek a generalization of the invariant interval Equation 4, between any two events $\mathscr{E}_a$ and $\mathscr{E}_b$ appropriate to the general case of Figure 6. The most obvious choice turns out to be the correct one; we draw on the basic assumption that space is isotropic and homogeneous. Thus all directions in space must be on an equal footing and physical properties must be unaffected by translations in space. The simplest generalization of Equation 4 satisfying these requirements is

$$(S_{ab})^2 = (ct_b - ct_a)^2 - (x_b^1 - x_a^1)^2 - (x_b^2 - x_a^2)^2 - (x_b^3 - x_a^3)^2. \tag{6}$$

Objectives 1 and 2   SAQ 2   Indicate briefly why the definition in Equation 6 satisfies the conditions of spatial isotropy and homogeneity.

Equation 6 is a useful definition because $(S_{ab})^2$ is a Lorentz invariant quantity in the general sense of Figure 6. This is so important that we repeat it and call it Theorem I:

> **Theorem I**
>
> The interval S_{ab} is a Lorentz invariant (in the sense of Figure 6); that is, if any two events, $\mathscr{E}_a$ and $\mathscr{E}_b$, are assigned coordinates $(ct_a, x_a^1, x_a^2, x_a^3)$, $(ct_b, x_b^1, x_b^2, x_b^3)$ by O and $(ct_a', x_a'^1, x_a'^2, x_a'^3)$, $(ct_b', x_b'^1, x_b'^2, x_b'^3)$ by O', then
>
> $$(S_{ab})^2 = (S_{ab}')^2.$$

You will not be assessed on the proof in the following passage.

Non-assessable ▼   To demonstrate Theorem I, let us first consider two events $\mathscr{E}_a$ and $\mathscr{E}_b$ that are
optional text   connected by a light pulse according to O. The pair $\mathscr{E}_a$ and $\mathscr{E}_b$ are then such, that observer O measures $(S_{ab})^2 = 0$. From Equation 6

$$c^2(t_b - t_a)^2 = (x_b^1 - x_a^1)^2 + (x_b^2 - x_a^2)^2 + (x_b^3 - x_a^3)^2$$

or

$$c(t_b - t_a) = \pm\sqrt{(x_b^1 - x_a^1)^2 + (x_b^2 - x_a^2)^2 + (x_b^3 - x_a^3)^2}.$$

Since the quantity on the right-hand side is just the three-dimensional spatial distance between the spatial locations of the two events, we see that this equation is just a three-dimensional generalization of Equation 3a. In other words, any pair of events whose coordinates satisfy the above equation are connected by a light signal according to O.

But, by postulate SR2, the propagation of light is the same for all inertial observers even if they are as in Figure 6. Thus observer O' will also observe that events $\mathscr{E}_a$ and $\mathscr{E}_b$ are connected by a light pulse. So observer O' can write

$$c(t_b' - t_a') = \pm\sqrt{(x_b'^1 - x_a'^1)^2 + (x_b'^2 - x_a'^2)^2 + (x_b'^3 - x_a'^3)^2}.$$

To put it another way, O' will conclude that his interval S_{ab}' will be zero if the interval S_{ab}, assigned to $\mathscr{E}_a$ and $\mathscr{E}_b$ by O, is zero. But by postulate SR1 all inertial observers are equivalent. Thus we may say that if S_{ab}' is zero, then S_{ab} must also be zero.

Next consider an *arbitrary* pair of events, $\mathscr{E}_a$ and $\mathscr{E}_b$. These events will not in general be connected by a light pulse. Observers O and O' will assign intervals squared of $(S_{ab})^2$ and $(S_{ab}')^2$ respectively to the event pair. These are not necessarily equal, but as we have shown above, if one is zero then so is the other. A simple connection between $(S_{ab})^2$ and $(S_{ab}')^2$ satisfying this condition would be

$$(S_{ab})^2 = A(\mathbf{V})(S_{ab}')^2.$$

Here we allow that A may depend on the relative velocity $\mathbf{V}$ between observers O and O'. But A *cannot* depend on any or all of the variables (ct, x^1, x^2, x^3) or (ct', x'^1, x'^2, x'^3), for if it did then certain points in spacetime would be singled out as more important than other points, contrary to the basic assumption that

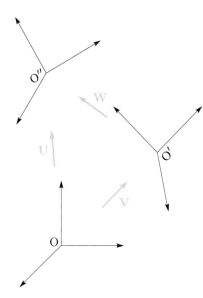

Figure 7   Three different (but equivalent) inertial frames.

spacetime is homogeneous (i.e. that all points (events) of spacetime have equal importance).

But we can go further than this: A can only depend upon the velocity $\mathbf{V}$ of O′ relative to O through its absolute value, $|\mathbf{V}|$. If this were not the case then the intervals S_{ab} and S'_{ab} would relate to one another in a way that depends on whether O and O′ were approaching or receding from one another; this is contrary to the idea of isotropy, that our 'distance' (i.e. interval) in spacetime should be independent of direction, as distances in three-dimensional Euclidean space are. We know that the distance between two points in ordinary Euclidean space is unaltered by rotations in any direction, so that the distance from P_A to P_B is the same as the distance from P_B to P_A. Thus,

$$A(\mathbf{V}) = A(|\mathbf{V}|)$$

and hence

$$(S_{ab})^2 = A(|\mathbf{V}|)(S'_{ab})^2. \tag{7}$$

So far we have considered just two frames, O′ moving at velocity $\mathbf{V}$ with respect to O as in Figure 6. Now consider a third inertial frame O″, moving with velocity $\mathbf{U}$ relative to O and with velocity $\mathbf{W}$ with respect to O′ (Figure 7). Then O″ will assign an interval squared, $(S''_{ab})^2$, to events $\mathscr{E}_a$ and $\mathscr{E}_b$ and we will have, as for Equation 7,

$$(S_{ab})^2 = A(|\mathbf{U}|)(S''_{ab})^2 \tag{8}$$
$$(S'_{ab})^2 = A(|\mathbf{W}|)(S''_{ab})^2. \tag{9}$$

We may eliminate $(S_{ab})^2$, $(S'_{ab})^2$ and $(S''_{ab})^2$ from Equations 7, 8 and 9 as follows. From Equations 7 and 8 we have:

$$A(|\mathbf{V}|)(S'_{ab})^2 = A(|\mathbf{U}|)(S''_{ab})^2.$$

Substituting for $(S'_{ab})^2/(S''_{ab})^2$ from Equation 9 gives:

$$A(|\mathbf{V}|)A(|\mathbf{W}|) = A(|\mathbf{U}|). \tag{10}$$

We now show that the only possibility of satisfying Equation 10 is that A be simply a constant and equal to unity.

To see this, write Equation 10 as

$$A(|\mathbf{W}|) = \frac{A(|\mathbf{U}|)}{A(|\mathbf{V}|)}. \tag{11}$$

Now the key point is that $|\mathbf{W}|$ actually depends on the relative directions of velocities $\mathbf{U}$ and $\mathbf{V}$. This is even true non-relativistically, for then the three velocities are related by simple vector addition, $\mathbf{U} = \mathbf{V} + \mathbf{W}$. So even in this non-relativistic case we would have

$$|\mathbf{W}|^2 = \mathbf{W} \cdot \mathbf{W} = |\mathbf{U}|^2 + |\mathbf{V}|^2 - 2\mathbf{U} \cdot \mathbf{V}$$

so that the magnitude of $\mathbf{W}$ does depend on the relative orientation of $\mathbf{U}$ and $\mathbf{V}$. Although, as you know from Section 3.3 of Unit 6, in special relativity velocities do *not* just add as $\mathbf{U} = \mathbf{V} + \mathbf{W}$, $|\mathbf{W}|$ certainly still depends on the relative orientation of $\mathbf{U}$ and $\mathbf{V}$. For example, if the direction of $\mathbf{V}$ were simply reversed (with $\mathbf{U}$ unchanged), then $\mathbf{W}$ would change both its direction and magnitude.

Now consider Equation 11. The left-hand side depends on $|\mathbf{W}|$ and, as we have just seen, therefore depends on the relative directions of $\mathbf{U}$ and $\mathbf{V}$. But the right-hand side of Equation 11 depends only on the *magnitudes* of $\mathbf{U}$ and $\mathbf{V}$ (i.e. on $|\mathbf{U}|$ and $|\mathbf{V}|$) and *not* on their directions. This is only possible if A is a constant independent of velocity; moreover $A = 1$ from Equation 11:

$$A = \frac{A}{A} = 1.$$

End of optional text   ▲

We have thus shown that the interval squared between any two events, *whether connected by a light pulse or not*, is a Lorentz invariant, completing our demonstration of Theorem I.

We shall be able to put the Lorentz invariance of the interval S_{ab} to deep and far ranging use throughout the remainder of this Unit.

3.3   A new notation

From the very outset of this Course we have dispensed with the familiar x, y, z labelling of spatial coordinates in favour of x^1, x^2, x^3. We said at the time that the advantage of doing this would become clearer at a later stage. This is where we begin to see the benefit. As we have often said, the old Galileo–Newton world-view of three-dimensional space plus a separate uniform absolute time, is replaced in special relativity by four-dimensional spacetime — the positions and times of events now get mixed up together between inertial observers. To emphasize this, we can adopt the convention of writing x^0 for ct (and x'^0 for ct', etc.). Note that ct has the dimensions of length.

The following self-assessment questions offer practice in the new notation. In addition, SAQ 4 asks you to verify from Theorem I that, for the special case of Figure 5, the transverse spatial coordinates of event $\mathscr{E}$ are indeed connected by the simple equations $x^2 = x'^2$ and $x^3 = x'^3$. These equations were first written down, without much proof, as Equations 23c and 23d of Unit 6, where we gave our derivation of the Lorentz transformation equations connecting coordinates (ct, x^1) and (ct', x'^1). Thus SAQ 4 clears up that final detail of Unit 6.

Objectives 1, 2 and 6   **SAQ 3**   Write down the Lorentz transformation equations for the case of Figure 5, using the new notation (you ought soon to be able to do this from memory). What is the invariant interval squared between events $\mathscr{E}_{\mathrm{a}}$ and $\mathscr{E}_{\mathrm{b}}$ for any observer such as O′ in Figure 6?

Objectives 1, 2 and 6   **SAQ 4**   In Section 3.1 you showed (SAQ 1) that if, for the special case of Figure 1, the ct (i.e. x^0) and x^1 coordinate transformation of an event pair $\mathscr{E}_{\mathrm{a}}$ and $\mathscr{E}_{\mathrm{b}}$ is given by Equations 1a and 1b, then the quantity $(ct_b - ct_a)^2 - (x^1_{\mathrm{b}} - x^1_{\mathrm{a}})^2$ is a Lorentz invariant. In the light of Theorem I, can you now show (using the new notation) that the equations $x^2 = x'^2$ and $x^3 = x'^3$ correctly transform the transverse spatial components of any event for the case of two inertial observers as in Figure 5?

Having introduced the new notation, we shall actually use it sparingly until we get to Section 6. It is important in the general relativity units.

4 Spacetime diagrams

4.1 Introduction

Let us now specialize to the case of inertial observers like O and O′ in Figure 1 whose x^1-axes are collinear, and let us consider events occurring on these x^1-axes. Owing to these assumptions, only two coordinates (ct, x), i.e. (x^0, x^1), are required to specify any event. Thus we have often been able to convey many of the concepts of relativity using *two*-dimensional spacetime diagrams.

We suppose O′ moves to the right with x-component of velocity $V \geqslant 0$ as observed by O. The lines of constant time and of constant position for O and for O′ are as shown in Figure 8. Note that a line of constant position is a possible world-line of a particle. A *line* of constant time is not a possible world-line of a particle — it would correspond to infinite speed. Nevertheless we can refer to lines of constant time on a spacetime diagram.

The essential properties of the Lorentz transformation are combined in the way lines of constant position and constant time according to one inertial observer appear on the spacetime diagram of another inertial observer. Let us now consider this problem.

An important property of the Lorentz transformation Equations 1a and 1b is their linearity. This ensures that straight line motion in one frame appears as straight line motion in another. Thus world-lines that are straight according to O′ will also be straight according to O. To see this, try the following self-assessment questions.

Objective 3   SAQ 5   A general straight world-line has the equation $ct' = Ax' + B$ according to O′, where A and B are arbitrary positive or negative constants. Sketch such a world-line on the usual (ct', x')-axes, indicating the relevance of A and of B. (Note that A must be $\geqslant 1$ for it to represent a possible world-line, $A = 1$ representing the case of light.)

It is clear that not only world-lines, but all straight lines on a spacetime diagram remain straight after a Lorentz transformation.

Objective 3   SAQ 6   Perhaps the simplest straight lines, according to O′, are his axes. The equation of the x'-axis, still according to O′, is simply $ct' = 0$. By using the Lorentz transformation, Equations 1a, 1b, and 1c, show that this line is observed by O as the straight line $ct = Vx/c$. Similarly, to O′, the equation of his ct'-axis is just $x' = 0$. Show that this straight line is observed by O as $ct = cx/V$. Sketch these two lines on axes (ct, x) and (ct', x'), indicating the slopes of these straight lines. Say briefly how these lines would appear to O if V/c were essentially zero (the non-relativistic limit).

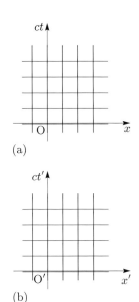

(a)

(b)

Figure 8 Lines of constant time (colour) and of constant position, for (a) observer O and (b) observer O'.

More generally, we may ask how straight lines parallel to the axes x' and ct' are registered by O on his axes. The horizontal (blue) lines in Figure 8(b) have the equation $ct' = a$, where a is any positive or negative constant. Using Equation 1b then gives the equation according to O:

$$\frac{1}{\gamma(V)}a = ct - \frac{V}{c}x$$

or

$$ct = \frac{V}{c}x + \frac{a}{\gamma(V)}. \tag{12}$$

This is (as expected!) a straight line with slope V/c.

Similarly, the vertical (black) lines (these are actually world-lines of stationary particles in O') in frame (ct', x') have the equation $x' = b$, where b is any constant. Using this in Equation 1a gives:

$$\frac{1}{\gamma(V)}b = x - \frac{V}{c}(ct)$$

or

$$ct = \frac{c}{V}x - \frac{c}{V}\frac{b}{\gamma(V)}.$$

This line has a slope of c/V. Without dwelling on all the details, the situation is summarized in Figure 9. Roughly speaking, lines of constant x' and lines of constant ct' for O' are squeezed towards the light cone (that is to say towards the line $x = ct$) for O.

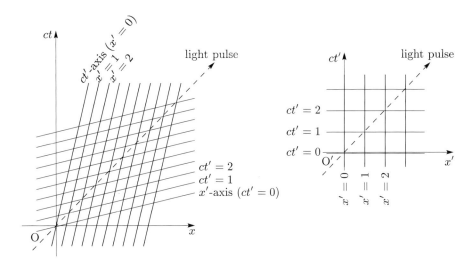

Figure 9 Lines of constant time and of constant position, according to O', are observed by O to be tilted and thus do not correspond (for O) to constant time or constant position.

It is possible to take the use of spacetime diagrams to great lengths, but for us their main value is that they can be used very speedily to represent general qualitative features.

4.2   Time-like and space-like intervals

We showed in Section 3 that the invariant interval squared $(S_{\mathrm{ab}})^2$, between any two events $\mathscr{E}_{\mathrm{a}}$ and $\mathscr{E}_{\mathrm{b}}$, is a Lorentz invariant quantity. Let us continue for the present to restrict ourselves to events occurring along the collinear

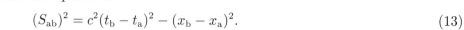

light cone

time-like intervals

space-like intervals

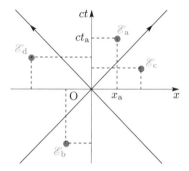

Figure 10 Events $\mathscr{E}_a$ and $\mathscr{E}_b$ are each separated from the origin by time-like intervals; events $\mathscr{E}_c$ and $\mathscr{E}_d$ are each separated from the origin by space-like intervals.

x- and x'-axes for two such observers, as in Figure 1. Then the invariant interval squared is

$$(S_{ab})^2 = c^2(t_b - t_a)^2 - (x_b - x_a)^2. \tag{13}$$

This quantity can be positive, negative or zero, allowing us to classify all event pairs into those for which $(S_{ab})^2 > 0$ or $(S_{ab})^2 < 0$ or $(S_{ab})^2 = 0$. When $(S_{ab})^2 = 0$, we say that $\mathscr{E}_a$ and $\mathscr{E}_b$ 'lie on the light cone' or are 'connected by a light pulse'; this second terminology, at least, is self-explanatory, for we should hardly need to remind you that $(S_{ab})^2 = 0$ means simply that $(x_b - x_a) = \pm c(t_b - t_a)$.

When $(S_{ab})^2$ is positive, $(ct_b - ct_a)^2 > (x_b - x_a)^2$ and we therefore say that the interval S_{ab} is *time-like*. Similarly, if $(S_{ab})^2$ is negative we say that the interval S_{ab} is *space-like*. Since in this latter case $(S_{ab})^2$ is negative, S_{ab} itself must be an imaginary number.

This classification of intervals (or pairs of events) is Lorentz invariant. By this we mean that if one inertial observer finds that the interval S_{ab} is, say, space-like, *then so will all other inertial observers*; this follows because all inertial observers will find the *same* value for $(S_{ab})^2$ and so agree that it is negative (space-like). Similarly, pairs of events which are separated by time-like intervals for one inertial observer will be time-like for all inertial observers.

Since this classification of intervals as space-like, time-like or connected by a light pulse is Lorentz invariant, we lose no generality by picking any particular inertial observer O as in Figures 10 and 11 and considering

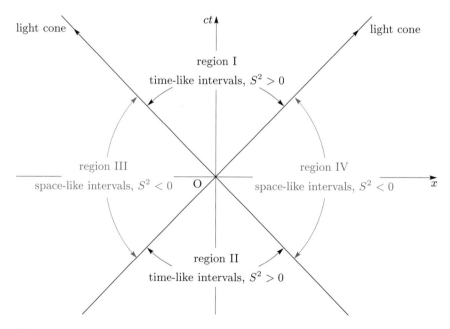

Figure 11 Classifying event pairs as time-like or space-like or as connected by a light pulse (lying on a light cone).

an arbitrary event with coordinates (ct, x) relative to the event $(0, 0)$ at the origin. In this case the interval squared is

$$(S)^2 = (ct)^2 - (x)^2.$$

For time-like intervals, $(S)^2 > 0$. This is true, for instance, for events like $\mathscr{E}_a$ and $\mathscr{E}_b$ (relative to the event $(0, 0)$ at the origin) in Figure 10, since the length of the side of the dashed rectangle corresponding to the ct-coordinate exceeds the length of the side corresponding to the

x-coordinate, giving $(ct_a)^2 > (x_a)^2$ and $(ct_b)^2 > (x_b)^2$. Thus we see that all events in regions I and II of Figure 11 are time-like with respect to the origin. Similarly, all events in regions III and IV of Figure 11 are space-like with respect to the event $(0,0)$.

Objective 2   **SAQ 7**   Argue briefly that all points (ct, x) in regions III and IV of Figure 11 do indeed correspond only to space-like intervals relative to the origin $(0,0)$.

Objective 2   **SAQ 8**   To emphasize that we have been classifying the interval squared, $(S_{ab})^2$, between pairs of events, consider an event $\mathscr{E}_a$ taking place anywhere on the spacetime diagram of observer O in Figure 12. Sketch in a figure like Figure 12 the regions of the (ct, x)-plane containing events that are space-like and time-like *with respect to $\mathscr{E}_a$*.

Figure 12   For SAQ 8.

Perhaps the reason for the term 'light cone' is unclear. If, for a moment, we restore one of the missing coordinates in Figure 11, say x^2, it will be perpendicular to the ct- and x^1-axes. (Unfortunately, we cannot put all four spacetime axes on a figure — an ongoing problem in the visualization of four-dimensional Minkowski spacetime.) Then the region with $c^2t^2 - (x^1)^2 - (x^2)^2 = 0$ will be the cone formed by rotating the lines designated 'light cone' in Figure 11 about the ct-axis.

4.3   Present, past and future

As you saw in the previous two Units, simultaneity is relative: two events that occur at the same time (but not the same place) according to one inertial observer may not occur simultaneously for another inertial observer who moves with respect to the first. This is clear in Figure 9, where (blue) lines of simultaneity for an observer O′, with equation $ct' = a$ (where a is any constant), are tilted according to observer O and so are not lines of simultaneity for O.

Although simultaneity is relative, we *can* make certain absolute statements concerning time because of the Lorentz invariant character of the invariant interval.

Consider an event pair with coordinates (ct, x) and $(0,0)$ in Figure 11. Any event with coordinates (ct, x) corresponds uniquely to some point in the plane of the spacetime diagram of observer O. The Lorentz transformation determines a unique point (ct', x') for every point (ct, x). In other words the Lorentz transformation is a mapping from the O-plane to the O′-plane. This is summarized in Figure 13 for the two observers of Figure 1 whose origins coincide at times $t = t' = 0$.

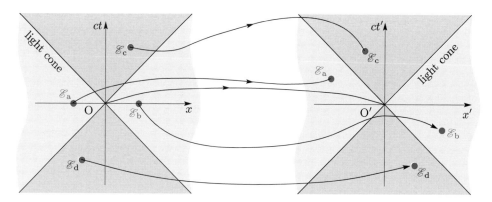

Figure 13   The Lorentz transformation is a mapping. Particular events in spacetime have different values of coordinates in different inertial frames.

All events whose coordinates lie in region I of the (ct, x)-plane of Figure 11 are time-like, with $(S)^2 > 0$. But $(S)^2$ is Lorentz invariant, so any other inertial observer O′ will agree that these events are time-like. Furthermore, O′ will agree that all these events lie in his region I, and not partly or wholly in his other time-like region, II. In fact, under the Lorentz mapping depicted in Figure 13, no event ever crosses the light cone. This follows from the Lorentz invariance of $(S)^2$; for if $(S)^2$ is positive for one inertial observer it is positive for all inertial observers, but to cross the light cone $(S)^2$ would have to change sign. *So this crossing cannot happen under a Lorentz transformation.*

Objective 2   **SAQ 9**   Argue briefly that, under the Lorentz transformation, events on the light cone of O must map to events on the light cone of O′.

Now choose an inertial observer, say O of Figure 11. We can say that, for O, any event with coordinates (ct, x) lying in region I occurs after the event $(0, 0)$ at his origin. This follows simply from the fact that $t > 0$ for all such events. Similarly all events (ct, x) lying in region II occur *before* the event $(0, 0)$ at his origin since $t < 0$ for all such events. But points in region I and *only* region I of observer O map (under the Lorentz transformation) into points in region I and only region I of any other inertial observer, say O′ of Figure 13. This argument applies equally to region II. Thus we have shown that events in region I are strictly future events (relative to the event at $(0, 0)$) according to all observers, and similarly events in region II are strictly past events (again relative to the event at $(0, 0)$). For these reasons region I of Figure 11 is often labelled **absolute future** *absolute future* of O and region II is often labelled *absolute past* of O.

Objective 3   **SAQ 10**   Indicate on a figure like Figure 12 which events occur in the absolute future and absolute past with respect to some arbitrary event $\mathscr{E}_\mathrm{a}$.

By way of summary thus far, we may say that in Figure 14 all events in region I of some event $\mathscr{E}_\mathrm{a}$ lie in the absolute future with respect to $\mathscr{E}_\mathrm{a}$ and are therefore agreed by all inertial observers to occur *after* $\mathcal{E}_a$, whereas all events in region II of the event $\mathcal{E}_a$ lie in the absolute past with respect to $\mathcal{E}_a$ and are therefore agreed by all inertial observers to occur *before* $\mathcal{E}_a$.

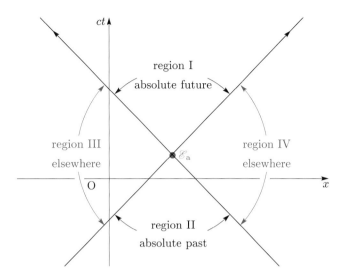

Figure 14   Event $\mathcal{E}_a$, its absolute past, its absolute future, and elsewhere.

causally related events

It is possible to refer to events occurring in regions I and II of any event $\mathcal{E}_a$ as being *causally related* to $\mathcal{E}_a$. Thus, any event in region I or on the upper light cone could be caused by a physical occurrence at $\mathcal{E}_a$; likewise, any event $\mathcal{E}_b$ in region II, or the lower light cone could be the cause of an occurrence at $\mathcal{E}_a$.

elsewhere

There are two regions we have not yet considered — the space-like regions III and IV which are labelled *elsewhere* in Figure 14. As argued above and as implied by Figure 13, events in regions III and IV will not cross the light cone under the Lorentz mapping from one observer to another. Thus events in regions III, say, will be agreed by all inertial observers to lie in their respective regions III, and similarly for region IV. But some events lying in region IV for observer O of Figure 11 will occur *before* the event at $(0,0)$ and other events of region IV will occur *after* the event at $(0,0)$. This is because some events in region IV of Figure 11 have $t < 0$ and others have $t > 0$. Since the only restriction is that events do not cross the light cone when mapped from one inertial observer to another, it is quite possible for an event in region IV to occur *before* the event $(0,0)$ for one observer and *after* the event $(0,0)$ for some other observer. It would all depend on whether the event mapped above or below the x-axis of the particular observer. In SAQ 11, this will be explicitly demonstrated. Thus, in regions III and IV the concepts of past and future are relative, with different observers disagreeing over the order of occurrence of events. As a special case, a thought-experiment in Sections 4.2 and 7 of Unit 5 demonstrated that two events that are simultaneous for one observer are not in general simultaneous for another observer moving with respect to the first.

More generally, in four-dimensional spacetime, Lorentz transformations can be found such that points in region III are 'rotated around' the light cone to region IV (and vice versa) without having to cross the light cone. The essential features of the present discussion are unchanged, however, by going to four-dimensional spacetime.

It is clear, then, that any event $\mathcal{E}_b$ occurring in regions III or IV of event $\mathcal{E}_a$ in Figure 14 cannot be causally related to $\mathcal{E}_a$. Many event pairs occur which are *not* causally related; if two observers disagree on the time order of events it is impossible to say that one event caused the other!

Objective 2

SAQ 11   Consider an event such as $\mathscr{E}_c$ of Figure 10. The invariant interval between $\mathscr{E}_c$ and the event $(0,0)$ is space-like; event $\mathscr{E}_c$ lies in region IV relative to the event $(0,0)$. Thus if the coordinates of $\mathscr{E}_c$ are (ct_c, x_c), then $(S)^2 = c^2 t_c^2 - (x_c)^2$ must be less than zero. According to observer O the time coordinate is positive, i.e. $t_c > 0$ (see Figure 10). Show that there exists another observer O$'$, such that $t'_c < 0$. You can do this by considering the Lorentz transformation Equation 1b.

5   Proper time

5.1   Introduction

One sometimes hears in popularized accounts that special relativity has nothing to say about phenomena that involve acceleration. In this Section we shall show that this is not true and in doing so introduce some ideas which will be of great importance when we study general relativity in Block 3.

In Section 5.2 we shall introduce the *instantaneous rest frame*; this, together with the properties of the invariant interval worked out in Sections 3 and 4, allows us to calculate the time elapsed on a clock moving with an arbitrarily accelerating particle (the proper time). The proper time can then be related to elapsed time in an *inertial* laboratory frame.

With these tools to hand we then return, in Section 5.3, to the so-called twin (or clock) 'paradox' introduced in its simplest form in Section 4.2 of Unit 6. We can now generalize to allow the out-and-back twin (Jack the Nimble) an arbitrary programme of acceleration and deceleration. The result is not trivial and carries a message about the geometry of spacetime itself (as we shall see in Section 2 of Units 10 & 11). In Section 5.4 we generalize the analysis of the relativistic Doppler effect given in Section 2.2 of Unit 6 to allow the moving source to have an arbitrary trajectory in three dimensions.

5.2   Preliminary ideas

Consider a particle moving along the x-axis of some inertial observer O. A typical world-line might be that in Figure 15. You can think of such a world-line as a continuous string of events. In Figure 15 two such events are labelled $\mathscr{E}_a$ and $\mathscr{E}_b$.

Objectives 2 and 3   SAQ 12   By drawing a light cone with origin at $\mathscr{E}_a$ in Figure 15, and assuming that no such material particle can travel faster than light, show that the event pair $(\mathscr{E}_b, \mathscr{E}_a)$ is separated by a *time-like* interval.

The result of SAQ 12 is extremely interesting, for it shows that, because any material particle must move at speeds less than c, it is possible to assign a sense or direction to any material particle's world-line on which all inertial observers can agree. This follows from the fact that all observers will agree that event $\mathscr{E}_b$ occurs *after* event $\mathscr{E}_a$, since $\mathscr{E}_b$ lies within the region of absolute future (region I of Section 4) with respect to $\mathscr{E}_a$. Similarly, by drawing a light cone with centre at $\mathscr{E}_b$ it follows that $\mathscr{E}_a$ lies in the region of absolute past with respect to $\mathscr{E}_b$. Now all other inertial observers who move with respect to O will observe the particle's world-line to have *different* shapes since the events on the world-line will be shifted about by the Lorentz transformation (see, for example, Figure 13). But from our analysis in Section 4.3 we can see that *they will all agree on the time order in which events on the world-line occur*.

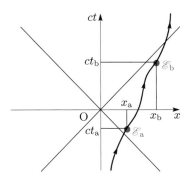

Figure 15 A typical world-line for a particle. Any material particle always moves at speeds less than c.

instantaneous rest frame

Now suppose that our particle carries a clock. Since we are considering a general particle world-line, as in Figure 15, we must accept that the particle may be accelerating at least some of the time. Therefore a rigid set of axes (x^1, x^2, x^3) (or in our present one-dimensional case, just x) attached to the particle would certainly *not* constitute an inertial system. One might thus suppose that no connection can be made between the time recorded on the particle's clock as it accelerates and moves along its world-line and the time recorded by some inertial observer like O in Figure 15.

However, if we assume that the actual process of acceleration doesn't affect the clock's inner workings (this is a good assumption for reasons to be mentioned shortly), a precise connection *can* be drawn by making use of a new construct: *an instantaneous rest frame*.

To approach the idea of an instantaneous rest frame, consider two trains moving along parallel tracks in the same direction. One train (train A) moves uniformly, whilst the other train (train B) accelerates steadily from a speed slower than that of A to a speed faster than that of A. There will be an interval of time, shorter or longer according to the acceleration of train B, during which coordinate systems fixed to each train will move along with no appreciable relative velocity. In other words, since B eventually accelerates past A there will an instant at which the two coordinate systems move at zero relative velocity, and for an arbitrarily small time interval about this instant their relative velocity will be arbitrarily small. *During this small interval we may say that system A is the instantaneous rest frame for system B.* An important point is that the instantaneous rest frame is an inertial frame since it is *not* accelerating. During the interval of time at which the two frames move together we may thus attempt to apply the methods of special relativity to analyse the physics in the accelerating frame in terms of the physics in the inertial frame. This involves a major assumption.

We assume that the acceleration itself has no effect on the physical phenomena taking place in the accelerating frame.

It is easy to think of everyday phenomena which *are* affected by acceleration. Drinking a glass of water is entirely different, for instance, in a rapidly accelerating train from what it is in a corresponding instantaneous rest frame. This is because forces come into play in the accelerating frame, that are absent from the corresponding inertial frame. Similarly, a mechanical clock carried in the accelerating frame would be subject to forces that might upset its time-keeping ability compared with a clock moving uniformly in the inertial frame. However, in principle, if we know exactly how acceleration affects the timekeeping we can always compensate for it exactly. Moreover, as will be discussed in the next Section, the decay of subatomic muons (a kind of heavy electrons) is quite unaffected by their centripetal acceleration around in a circle at something like 10^{18} times the acceleration of gravity ($9.8 \, \mathrm{m \, s^{-2}}$)!

It seems then that atomic and subatomic processes of Nature are, to a good approximation at least, unaffected by acceleration. We therefore expect that with the help of the idea of the instantaneous rest frame we can isolate the intrinsic properties of space and time from the superficial effects of acceleration on many mechanical devices.

Consider two infinitesimally separated events (ct, x) and $(ct + c\Delta t, x + \Delta x)$ on a particle's world-line, as in Figure 16. We suppose that Δx and Δt are

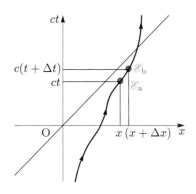

Figure 16 $\mathscr{E}_a$ and $\mathscr{E}_b$ are an infinitesimally separated pair of events on the particle's world-line.

infinitesimally small. If we think of an *inertial* frame O′ that moves (in the case of Figure 16) to the right in frame O with velocity $\Delta x / \Delta t$, then the particle will, for a short time at least, be at rest relative to O′. Since even during the short time Δt the particle may be accelerating, there may be a small correction to this, but this correction will *become zero* (at time t) as Δx and Δt are allowed to approach zero. A *whole series* of such instantaneous rest frames may, if necessary, be mentally called into play for every point along the particle's world-line. For the moment, however, we need only consider the two infinitesimally separated events of Figure 16, assuming that anything that applies to them applies to any such events.

The frame O of Figure 16 and the instantaneous rest frame moving with the particle near event (ct, x) of Figure 16 are both inertial frames. Thus they will assign the *same* value to the Lorentz invariant interval between the infinitesimally separated events $\mathscr{E}_a$ and $\mathscr{E}_b$ of Figure 16. Since $\mathscr{E}_a$ and $\mathscr{E}_b$ are infinitesimally separated, we shall write $(\Delta S)^2$ for the interval squared. According to O this is

$$(\Delta S)^2 = c^2(\Delta t)^2 - (\Delta x)^2. \tag{14}$$

According to the instantaneous rest observer, whom we shall call O′, this is

$$(\Delta S)^2 = c^2(\Delta t')^2 - (\Delta x')^2. \tag{15}$$

Note by writing $(\Delta S)^2$ in both Equations 14 and 15 we have used the fact that the interval is a Lorentz invariant quantity.

Now to O′, who is moving along with the particle at the instant under consideration so that the particle is at rest in his frame, the spatial separation $\Delta x'$ between $\mathscr{E}_a$ and $\mathscr{E}_b$ becomes zero (to an approximation that becomes exact as $\mathscr{E}_b$ and $\mathscr{E}_a$ become increasingly close). Thus Equation 15 gives

$$(\Delta S)^2 = c^2(\Delta t')^2. \tag{16}$$

Objectives 2 and 3   **SAQ 13**   By combining Equations 14 and 16 show that

$$\Delta t' = \Delta t \sqrt{1 - \frac{1}{c^2}\left(\frac{\Delta x}{\Delta t}\right)^2} \tag{17}$$

thus giving for the (time-like) interval the value

$$\Delta S = c\,\Delta t \sqrt{1 - \frac{1}{c^2}\left(\frac{\Delta x}{\Delta t}\right)^2}. \tag{18}$$

Now examine Equations 17 and 18. In these equations, Δt is the time between two infinitesimally separated events on the particle's world-line according to O, and $\Delta t'$ is the corresponding infinitesimal time increment according to O′ who is moving along with the particle at the instant under consideration. But since the particle is at rest with respect to O′ at that instant, we can also say that $\Delta t'$ is the infinitesimal time increment *recorded by the clock attached to the particle* between the two events. And, by Equation 16, the invariant interval is $\Delta S = c\,\Delta t'$, which gives

$$\Delta t' = \frac{\Delta S}{c}.$$

proper time

This time increment $\Delta t'$, elapsed on a clock moving with the particle, is sufficiently important to warrant a name of its own; it is called the *proper time* between events $\mathscr{E}_a$ and $\mathscr{E}_b$. We shall write $\Delta \tau$ for the proper time between infinitesimally separated events. Thus we can make the following statement:

The increment of time $\Delta \tau$, between any two infinitesimally separated events on a particle's world-line as recorded by a clock (an atomic clock, say) moving with the particle, is equal to the invariant interval between these two events divided by c:

$$\Delta \tau = \frac{\Delta S}{c}. \tag{19}$$

$\Delta \tau$ is called the *proper time* interval between $\mathscr{E}_a$ and $\mathscr{E}_b$.

This proper time increment can be calculated by any inertial observer O, by means of Equation 17, noting that, as $\Delta t = t_b - t_a$ goes to zero, $\Delta x / \Delta t$ becomes the particle's velocity, $u(t)$ say, evaluated at some time between t_b and t_a. This gives, with Equation 18,

$$\Delta S = c \, \Delta t \, \sqrt{1 - \left(\frac{u(t)}{c}\right)^2}. \tag{20}$$

It is important to remember that $u(t)$ is the velocity of the particle as measured by O, as in Figure 16.

Let us summarize briefly the argument to this point. We have considered an arbitrary world-line for a material particle, as in Figures 15 and 16. By the construct of the instantaneous rest frame we have shown that for two infinitesimally separated points on the particle's world-line, such as $\mathscr{E}_a$ and $\mathscr{E}_b$ of Figure 16, the time increment recorded between $\mathscr{E}_a$ and $\mathscr{E}_b$ by a clock moving with the particle is given as $\Delta S / c$, where ΔS is the (infinitesimal) invariant interval connecting $\mathscr{E}_a$ and $\mathscr{E}_b$. If the motion of the particle is observed from any inertial frame, such as O in Figure 16, then this same (infinitesimal) invariant interval is given by Equation 20, where $u(t)$ is the particle's velocity (as observed by O) at the time under consideration. Since $\mathscr{E}_a$ and $\mathscr{E}_b$ are infinitesimally close it is immaterial whether $u(t)$ is evaluated at time t_a or t_b, or indeed at some value in between. The time elapsed on the clock as it moves from $\mathscr{E}_a$ to $\mathscr{E}_b$ with the particle is called the proper time increment, $\Delta \tau$. Thus we have

$$(\Delta \tau)_{ab} = \frac{1}{c}(\Delta S)_{ab}. \tag{21}$$

Objective 5 **SAQ 14** Write $(\Delta S)^2$ in terms of the new notation involving x^0.

Objectives 1, 2 & 3 **SAQ 15** Describe briefly the main steps leading to the concept of proper time, starting with the properties of the invariant interval worked out in Section 3.1. Include in your list any essential equations.

5.3 The significance of proper time; the twin 'paradox' revisited (Theorem II)

The world-line of a particle may be considered to be composed of a large number of infinitesimal displacements, as included in Figure 17. As the particle moves from event $\mathscr{E}_c$ to event $\mathscr{E}_d$ it will, in general, be accelerating or decelerating. By splitting the time interval $(t_d - t_c)$ (according to O) into small intervals, we may use the methods of Section 5.2 to cope with this case. Since we may think of the velocity as virtually constant throughout each small increment of time Δt, we can imagine a whole series of instantaneous rest frames each at rest with respect to the particle during its respective time interval.

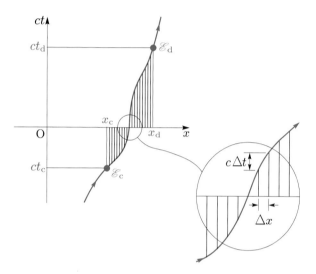

Figure 17 The motion of a material particle along its world-line can be imagined as composed of small steps.

If the particle carries a clock, then the time elapsed on the clock as its motion carries it from $\mathscr{E}_c$ to $\mathscr{E}_d$ will be the sum of the proper time intervals $\Delta\tau$, for each section of the world-line. This time elapsed on the particle's clock is just the proper time τ_{cd}, between events $\mathscr{E}_c$ and $\mathscr{E}_d$ along the world-line of the clock. As we shall see from Theorem II, a *different* world-line joining $\mathscr{E}_c$ and $\mathscr{E}_d$ would, in general, correspond to a *different* elapsed proper time.

Observer O can calculate τ_{cd} provided he can ascertain the particle's velocity $u(t)$, at all times t lying between t_c and t_d; for by Equations 20 and 21 it is known that

$$\tau_{cd} = \text{ the sum from } t_c \text{ to } t_d \text{ of terms like } \Delta t \sqrt{1 - \left(\frac{u(t)}{c}\right)^2}. \qquad (22)$$

The proper time from $\mathscr{E}_c$ to $\mathscr{E}_d$ of Figure 17 is physically very important. In particular, the proper time on the world-line from $\mathscr{E}_c$ to $\mathscr{E}_d$ *is the time recorded on the particle's clock as it moves between the two events*. Since the particle's clock will advance continuously (its hand always moving clockwise as seen by a data taker also moving with the particle), and since (by Equation 19) the proper time is a Lorentz invariant quantity, we may unambiguously label every point (event) on the particle's world-line with respect to any initial reference point ($\mathscr{E}_c$ say) by means of the proper time. This ability of proper time to give a unique label to points on a world-line

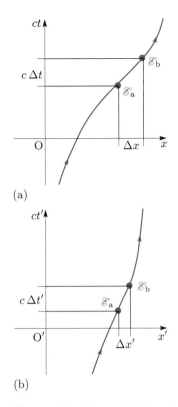

(a)

(b)

Figure 18   The world-line of the particle, (a) according to O and (b) according to O'.

will remain very important for us when we study general relativity in Units 10 & 11.

In detail, suppose two inertial observers, O and O', have recorded the world-line of a particle as in Figures 18(a) and 18(b). Both O and O' have recorded the *same* world-line but, because simultaneity is *relative*, their plotted trajectories look different, even though they used identical clocks and metre sticks. O has established the position x, of the particle as a function of t, whereas O' knows x' as a function of t'. Thus O can find the particle's velocity, $\Delta x/\Delta t$, as a function of t, and O' knows the corresponding velocity, $\Delta x'/\Delta t'$, as a function of t'. By means of Equation 22, both observers can find the *same* (invariant) proper time elapsed between events $\mathscr{E}_a$ and $\mathscr{E}_b$ of Figures 18(a) and 18(b). In this way, O finds the time elapsed as a function of proper time, namely $t = A(\tau)$. Similarly, O' finds $t' = B(\tau)$.

Finally, O knows the particle's world-line (Figure 18(a)), so he knows x as a function of t. But, since t is a function of τ, he has the dependence of x upon τ, namely $x = C(\tau)$. Similarly O' can establish x' as a function of τ, namely $x' = D(\tau)$.

In summary, writing x^0 for ct we can say that O knows the τ-dependence of both x^0 and x^1 along the particle's world-line, namely $x^0(\tau)$ and $x^1(\tau)$, and that any other observer O' also can be assumed to know $x'^0(\tau)$ and $x'^1(\tau)$.

Objectives 2, 3 & 9   SAQ 16   Argue briefly that this unique labelling of the motion of a particle on its world-line is consistent with our previous discussion of past and future, especially as summarized in the paragraph immediately preceding Figure 14.

We can now deal with the so-called twin (or clock) 'paradox' in its most general form. It is an immediate consequence of an important result we shall label as Theorem II:

Theorem II

The proper time elapsed between any two events on a material particle's world-line is equal to or less than the corresponding time recorded between these two events by any inertial observer not moving with the particle. If the particle returns to its point of departure, then its elapsed proper time is strictly less than that for an inertial observer who has remained at that point in space.

The proof of Theorem II is easy. For according to Equation 22, the proper time elapsed during every increment of the world-line equals the inertial observer's elapsed time, Δt, multiplied by a factor equal to or less than unity. This factor can only equal unity if $u(t)$, the particle's velocity relative to the observer, becomes zero. But even if this *is* the case during one instant of time, it cannot remain so if the particle accelerates in any way, for the inertial observer, being inertial, does not accelerate. In particular (this is the second half of the theorem), if the particle leaves an inertial observer and returns later, it *must* have accelerated. Thus, since its clock reads the proper time, its elapsed time from leaving to returning to the inertial observer (as measured on the particle's own clock) will be less than the time elapsed on the observer's clock; if one twin moves out and back with this particle then, on return, his biological clock (which is, after all, governed by atomic processes) will have aged less than his stay-at-home brother.

Far from being a paradox, *Theorem II is a fundamental statement about the properties of space and time.* In Section 9 of Units 10 & 11 it will lead to a fundamental *geometrical* statement about the motion of free particles in spacetime.

Theorem II makes a strong statement that cries out for experimental testing. How could it be tested?

It will be many years before twin-paradox experiments can be performed on real pairs of identical human twins; rockets are just too slow for $\sqrt{1 - u^2/c^2}$ to differ sufficiently from unity to result in recognizably different ageing. But u/c can approach unity for elementary particles in particle accelerators. You saw in Section 3.2 of Unit 6, when we first discussed time dilation, how the internal 'self-destruct' clock of muons, governing their decay into electrons and other particles, provides a test for time dilation, and explains the apparent long life of muons created in cosmic ray showers. So we have one of the essentials of a twin paradox test: a particle, carrying its own clock, which is identical to particles which can be kept 'at home' in the lab. Instead of studying the decay of muons generated in the upper atmosphere, we can study muons in something more like the twin situation in which they return to to the starting point after a programme of acceleration. In fact, the muons return to the starting point thousands of times, circulating in a large circular storage ring, kept in orbit by the centripetal acceleration due to strong magnets. Along the world-line of a muon, the elapsed proper time governs its decay, but as we show below, the lifetime as seen from observers at rest near the storage ring will be $\gamma(u)$ times greater, where u is the constant speed of the muons around the storage ring. The idea is to compare the lifetime of the circulating muons with that of identical muons at rest.

J. Bailey *et al.* (1977), Measurements of relativistic time dilation for positive and negative muons in a circular orbit. *Nature, London,* **268**, 301–304.

This was the measurement carried out in the CERN Muon Storage Ring by J. Bailey and co-workers. In this experiment a large number of muons (because radioactive decay is a statistical process; the half-life is the time over which half of the particles decay) were stored in a circular trajectory about 14 metres in diameter. When at rest, muons decay after an average time of $\sim 2.2 \times 10^{-6}$ s (Unit 6, Section 3.2).

The *half-life* is the time in which, on average, half the particles will decay. The *mean life* is the average lifetime of a particle. They are related by
half-life = 0.693 × mean life, in which 0.693 = ln 2.

Each time around the ring, a muon leaves its starting point and returns, moving with constant speed relative to the laboratory frame but undergoing a constant centripetal acceleration (i.e. the speed stays the same but the direction changes). Although Equation 22 strictly applies to motion along one axis, it is not difficult to see (from Section 5.4) that in the general case of motion in four-dimensional spacetime, with a steady speed u relative to the laboratory frame, the proper time of a muon decay, τ, is related to the decay time T observed in the laboratory, by

$$\tau = T\sqrt{1 - \left(\frac{u}{c}\right)^2} = T/\gamma(u). \tag{23}$$

In the experiment, the measured value of T for negatively charged muons was $64.368 \pm 0.029 \times 10^{-6}$ s, where their speed was such that $\gamma(u)$ (defined in Equation 1c) was 29.33; τ, the at-rest lifetime was measured to be $2.196\,66 \pm 0.002\,0 \times 10^{-6}$ s.

Objectives 4 and 9   SAQ 17   Did experiment verify theory in this case?

In 1971, Hafele and Keating carried out an even more direct 'twin paradox' experiment. The twins in this case were identical atomic clocks, and the idea was to take an atomic clock around the world in a commercial airliner, and compare the elapsed time with that of a stationary clock. In fact, it was somewhat more complex than this, since another source of time dilation (due to gravity, to be discussed in Unit 9) was active. For this reason, two atomic clocks were taken in opposite directions around the world and compared, enabling the gravitational time dilation effect, as well as the rotation of the Earth, to be disentangled from the time dilation due to the speed. In that experiment, and more recent and more precise versions of the experiment, the differential ageing of twins, even if they were twin clocks, occurred exactly in accord with theory.

5.4   The relativistic Doppler shift and time dilation in four-dimensional spacetime

It is obvious that, in general, particles move in arbitrary directions in three-dimensional space, and not just along an x-axis. In this Section, we shall use the results of Section 3.2 to derive the generalizations of (i) the time dilation expression (Equation 26 of Unit 6) and, (ii) the relativistic Doppler effect (Equation 5 of Unit 6). You will recall that, for motion in one dimension, these two effects were related to one another in Section 4.1 of Unit 6; similar reasoning is followed here.

5.4.1   Time dilation

Suppose a particle moves in some arbitrary direction in three-dimensional space. An inertial observer O ascribes spacetime coordinates (ct, x^1, x^2, x^3) to the four-dimensional world-line of the particle.

Consider two ticks of a clock carried by this particle. If, according to O, these two events are separated by infinitesimal increments $(c\,\Delta t, \Delta x^1, \Delta x^2, \Delta x^3)$, then according to Equation 6 the invariant interval is given by

$$(\Delta S)^2 = c^2(\Delta t)^2 - \left\{(\Delta x^1)^2 + (\Delta x^2)^2 + (\Delta x^3)^2\right\}.$$

But the proper time interval between these ticks is $\Delta \tau = \Delta S/c$. Thus

$$(\Delta \tau)^2 = (\Delta t)^2 - \frac{1}{c^2}\left\{(\Delta x^1)^2 + (\Delta x^2)^2 + (\Delta x^3)^2\right\}.$$

If we rearrange this and take the square root we find,

$$\Delta \tau = \Delta t \sqrt{1 - \frac{1}{c^2}\left\{\left(\frac{\Delta x^1}{\Delta t}\right)^2 + \left(\frac{\Delta x^2}{\Delta t}\right)^2 + \left(\frac{\Delta x^3}{\Delta t}\right)^2\right\}}.$$

Finally, since $\Delta x^{\mathbf{1}}/\Delta t$ is the **1**-component of the particle's velocity in the frame of O (and similarly for the **2**- and **3**-directions), we get

$$\Delta \tau = \Delta t \sqrt{1 - \left(\frac{u}{c}\right)^2} = \Delta t / \gamma(u) \tag{24}$$

where u is the speed of the particle in three dimensions. Equation 24 is the obvious generalization of the one-dimensional time dilation equation (Equation 26 of Unit 6).

Objectives 1 and 3   SAQ 18   Describe in words the meaning of Equation 24.

5.4.2   Relativistic Doppler effect in three dimensions

The situation is illustrated in Figure 19.

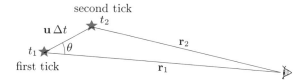

Figure 19   Between two ticks, the clock moves by a displacement $\mathbf{u} \, \Delta t$.

Imagine an arbitrarily moving clock that emits a pulse of light towards an inertial data taker every time it ticks. Consider two consecutive ticks. The proper time between ticks (that is, the time between ticks according to the moving clock) is $\Delta \tau$. Let the time between ticks as observed in the frame of the data taker be $\Delta t = t_2 - t_1$ (t_1 and t_2 being the times that the ticks *happen* in the observer's frame, not the times he sees them happen). During this time Δt the clock moves closer to (or farther from) the observer at a rate $u \cos \theta$, where u is the speed of the clock which is travelling at an angle θ to the line of sight of the observer. Thus in time Δt the clock moves closer to the data taker by an amount $(u \cos \theta) \, \Delta t$. Since the second pulse is emitted that much closer to the data taker we can say

$$\text{time between received pulses} = \Delta t - \frac{(u \cos \theta) \, \Delta t}{c}.$$

But the time Δt between ticks according to the observing system is related to the time $\Delta \tau$ between ticks according to the moving clock by the time dilation formula, Equation 24. Thus,

$$\text{time between received pulses} = \frac{\Delta \tau}{\sqrt{1 - \left(\frac{u}{c}\right)^2}} \left(1 - \frac{u}{c} \cos \theta\right)$$

$$= \gamma(u) \left(1 - \frac{u}{c} \cos \theta\right) \Delta \tau. \tag{25}$$

This is the relativistic Doppler shift appropriate to arbitrary motion in three dimensions.

Objective 4   SAQ 19   Show that the right-hand side of Equation 25 reduces to the result in Section 4.1 of Unit 6, namely

$$\Delta\tau\sqrt{\frac{1-u/c}{1+u/c}}$$

when the source is moving along the line of sight towards the observer.

Objective 4   SAQ 20   Suppose the source is continuously emitting monochromatic radiation of frequency f_0. Give an expression, using Equation 25, for the received frequency, f.

When the source is moving transversely to the observer, i.e. $\cos\theta = 0$, the Doppler expression reduces to:

$$\text{time between received pulses} = \gamma(u)\,\tau.$$

transverse Doppler shift   This is the *transverse Doppler shift*. It is essentially an effect of relativistic time dilation. Though a small effect, it has been observed, and will be met again in Unit 9.

6 Linear momentum and energy in special relativity

6.1 Introduction

You have seen that, for any situation where all the relevant speeds are much less than c, Newtonian concepts of space and time work and time is effectively absolute. We speak of the 'non-relativistic limit'. In the same way, we expect that new concepts introduced into relativistic mechanics should, in the non-relativistic limit, correspond to Newtonian energy, momentum, force, and so forth.

Here, we shall only generalize linear momentum and energy. Nevertheless, this will allow us to discover certain striking departures from Newtonian mechanics for swiftly moving bodies.

6.2 The definition of linear momentum (a thought-experiment)

In Newtonian mechanics the linear momentum of a particle becomes important when forces act. For example, if there is a closed system of particles interacting with one another but acted on by no external forces, then the total linear momentum of all the bodies is *conserved*. A special case of this is the collision between two particles such as billiard balls. Before and after the collision the balls move freely, but during the encounter complicated processes of compression and relaxation occur in the substance of the balls. We can say, however, that

total linear momentum before collision = total linear momentum after collision.

This statement is true even if the collision is inelastic, i.e. even if in the collision some of the initial kinetic energy is lost to heat etc.

Objective 7 SAQ 21 Is momentum conserved when I throw a ball at a wall and it bounces back at me?

In Newtonian mechanics, the linear momentum of a particle of mass m and velocity $\mathbf{v}$ is $m\mathbf{v}$. We shall now consider an especially simple collision to illustrate how relativity requires this definition to be altered in order that linear momentum should continue to be conserved in collisions.

Momentum is to be defined so as to be conserved in collisions; so therefore let us apply momentum conservation to a collision chosen so that symmetry makes the outcome easy to see. Such a collision is depicted in Figure 20. It involves a glancing collision between two identical particles, such as billiard balls, in the inertial laboratory frame. By 'glancing' we mean that, although the velocities of particles A and A' along the laboratory x^1-axis may be very large, their transverse components of velocity (along the x^2-axis) are very small. Notice that we have oriented the x^1- and x^2-axes so as to emphasize the symmetry inherent in the

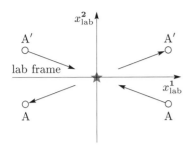

Figure 20 A glancing elastic collision in the laboratory frame. For clarity, the collision shown is not particularly glancing.

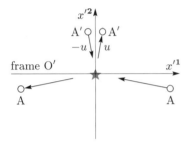

Figure 21 O' moves with A' along x_{lab}^1. The figure shows a small motion of A' in the x'^1-direction just for clarity.

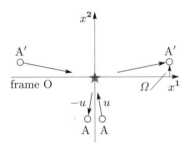

Figure 22 O moves with A along the negative x_{lab}^1-direction.

problem. As implied in Figure 20 the collision is such that (in the laboratory frame) the initial speeds of the two particles are equal. The collision is also *elastic* i.e. no kinetic energy is converted into heat or other forms of energy. Since the initial speeds of A and A' are set up to be equal and since their masses are equal, it then follows from the symmetry of the problem that the only effect of the collision can be a symmetric change in direction, and this too is shown in Figure 20.

To arrive at a definition of linear momentum we must make further assumptions. We assume that the linear momentum of a particle has the form $\mathbf{p} = mf(v)\mathbf{v}$, where m is the particle's mass and $\mathbf{v}$ is its velocity. The function f will be assumed to depend on the speed $v = |\mathbf{v}|$ of the particle. Lastly, we insist that the collision of Figure 20 conserves linear momentum. This conservation will then dictate the function $f(v)$ in our assumed form for linear momentum.

To exploit the great symmetry of Figure 20, let us consider the collision from two different inertial frames.

We choose the first frame O', to move to the right relative to the laboratory frame at such a speed that it cancels out the component of the velocity of A' along the x_{lab}^1-direction. In this frame, A will move even faster from right to left. This is shown in Figure 21 where we have tried to indicate that in O', particle A' has no x'^1-component of velocity but simply moves slowly down and then up the x'^2-axis with speed $|u|$; 'slowly', because the experiment has been set up to be glancing, with only small transverse velocities for A and A'.

Similarly, we consider the collision from a second inertial frame O, moving from right to left in the laboratory frame at such a speed that it cancels out the x^1-component of the velocity of A. In this frame, A moves slowly up and then down with the same speed $|u|$ as the speed of A' in frame O'. This is shown in Figure 22.

Now frame O' moves to the right, relative to the laboratory frame, at some velocity $\mathbf{V}$, and frame O moves to the left at velocity $-\mathbf{V}$. But, more to the point, frame O' moves at some possibly large velocity $\mathbf{W}$, relative to O.

Since frames O and O' are inertial frames, we may use the velocity addition equation developed in Unit 6 to relate the transverse component of the velocity of A' after the collision in frame O (indicated as Ω in Figure 22) to the transverse velocity of A' after the collision in frame O' (indicated as u in Figure 21).

Objective 7   SAQ 22   Show, from Equation 31 of Unit 6, that

$$\Omega = \frac{u}{\gamma(W)}. \tag{26}$$

Our premise is that momentum is conserved in the collision. The laboratory frame has been chosen so that the net (i.e. total) transverse (x^2)-component of momentum is zero before and after the collision. Similarly the transverse component of momentum in frames O' and O must be zero. It should be clear from Figure 22 that if there were a net transverse momentum before the collision then it would reverse direction after the collision so the only possibility is that there is zero net transverse momentum both before and after the collision.

We are analysing a glancing collision so u and Ω are effectively small. Thus, to an approximation increasingly good as $u, \Omega \longrightarrow 0$, the x^1-component of the velocity of A′ in frame O is just W, the velocity of frame O′ relative to frame O. To the same approximation, the statement that the net transverse momentum is zero in frame O (see Figure 22) is written:

transverse component of momentum of A

= transverse component of momentum of A′

or, evaluating $p^2 = m f(v) v^2$ for A and A′,

$$m f(u)\, u = m f(W)\, \Omega.$$

But u is arbitrarily small, so the transverse component of momentum is Newtonian, with $f(u) \approx 1$. Using this, together with Equation 26 for Ω, gives:

$$f(W) \approx \frac{u}{\Omega} = \gamma(W).$$

As u approaches zero this becomes exact. Notice that W is not necessarily small.

linear momentum

We are thus led to *define* the *linear momentum* of a free particle whose mass is m by

$$\mathbf{p} = m\,\gamma(v)\,\mathbf{v} \tag{27a}$$

where, as usual,

$$\gamma(v) = \frac{1}{\sqrt{1 - \left(\dfrac{v}{c}\right)^2}}. \tag{27b}$$

Although we have motivated this definition of linear momentum by considering a particular class of collision, it has been found to give consistent results for an enormous range of elastic and inelastic collisions between particles. Moreover, it leads to a form of relativistic mechanics which has a crucial property that Newton's mechanics does not have: it can be written in the same form in all inertial frames.

The new definition of momentum differs from the Newtonian definition by the factor γ. How is this to be interpreted? In some books (mostly older ones) the factor γm is referred to as the 'relativistic mass' whereas m is referred to as the 'rest mass'. Of course, m is the value taken by γm as $\gamma \longrightarrow 1$ as $v \longrightarrow 0$ with the particle at rest. We follow the more modern convention according to which m is simply the mass of the particle and we do not *require* the term 'rest mass'. The factor γ means that, to an extent that increases with v, the momentum of a particle of given mass and velocity, is larger than that calculated according to the Newtonian expression. Notice that the mass m of a body is an intrinsic property of the body and by definition is the same in all inertial frames; it is therefore a Lorentz invariant.

m is the same in all inertial frames by definition.

The rest of Unit 7 is devoted to the consequences of the definition in Equation 27. Equations 27a and 27b are the vital results you should remember from Section 6.2.

Objectives 1 and 7   **SAQ 23**   What would you say are the major assumptions in the analysis leading to the definition in Equation 27?

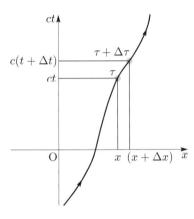

Figure 23   World-line of a particle moving along the x-axis.

6.3 Space and time; momentum and energy (Theorem III)

In Section 6.2 we defined the momentum of a body of mass m and velocity $\mathbf{v}$ as $\mathbf{p} = \gamma(v)m\mathbf{v}$, motivating this definition by requiring that momentum be conserved in a glancing collision. We now show that this definition of momentum leads to a new definition of the energy of a particle.

For convenience, consider one spatial coordinate, the x^1-axis (alias x-axis). Then we can portray a particle's world-line on a two-dimensional spacetime diagram as in Figure 23. We emphasized in Section 5.3 that each point of the motion of a particle along its world-line can be labelled by the proper time elapsed; and since proper time is a Lorentz invariant, this labelling is itself Lorentz invariant. Since the position of the particle, as it follows its world-line, is a function of the proper time elapsed, the observer O of Figure 23 would assign $(x^0(\tau), x^1(\tau)) = (ct(\tau),\ x(\tau))$ as the spacetime position of the particle at proper time τ. Any other inertial observer O′ would assign position $(x'^0(\tau), x'^1(\tau)) = (ct'(\tau),\ x'(\tau))$.

The implications of this are great. In the case we are considering of motion along the x-axis (i.e. x is alias for x^1), the linear momentum is $m\,\gamma(v)\Delta x/\Delta t$, since $\Delta x/\Delta t$ is the particle's velocity according to O. (Strictly, the velocity is dx/dt, but it is useful to keep the intervals finite as we manipulate them, with the possibility of going to the limit when required. We refer to quantities like $\Delta\tau$ as infinitesimals since we can make them as small as necessary for any given level of accuracy.)

Expressing the infinitesimal invariant interval in the customary way (see, for instance, Equation 13), we have

$$(\Delta S)^2 = c^2(\Delta t)^2 - (\Delta x)^2.$$

Dividing by $c^2(\Delta t)^2$ and replacing ΔS by $c\,\Delta\tau$ (where $\Delta\tau$ is infinitesimal invariant proper time interval) gives

$$\left(\frac{\Delta\tau}{\Delta t}\right)^2 = 1 - \frac{1}{c^2}\left(\frac{\Delta x}{\Delta t}\right)^2.$$

Recognizing $\Delta x/\Delta t$ as the particle's velocity, we have therefore shown that

$$\left(\frac{\Delta\tau}{\Delta t}\right)^2 = 1 - \frac{v^2}{c^2} = \left(\frac{1}{\gamma(v)}\right)^2$$

or

$$\frac{\Delta t}{\Delta\tau} = \gamma(v). \tag{28}$$

This is a generalization of Equation 24 which was derived for uniform motion; it provides a new way of writing the linear momentum of the particle (recalling that x was alias for x^1):

$$p^1 = m\,\gamma(v)\left(\frac{\Delta x^1}{\Delta t}\right) = m\left(\frac{\Delta t}{\Delta\tau}\right)\left(\frac{\Delta x^1}{\Delta t}\right).$$

The factor Δt may be cancelled to give

$$p^1(\tau) = m\frac{\Delta x^1(\tau)}{\Delta \tau}. \tag{29}$$

In writing this equation we have reintroduced the superscript notation to emphasize that we are considering motion along the **1**-direction, and we have indicated that, as the particle moves along its world-line, its momentum and position can be considered functions of the proper time, τ.

Objectives 1, 5 & 7   **SAQ 24**   What happened to the time t? Cannot p^1 and x^1 be considered (by observer O) as functions of t?

Remember from Section 6.2 that mass is Lorentz invariant.

In Equation 29, p^1 is the product of a Lorentz invariant quantity, namely $m/\Delta\tau$, and an increment of the **1**-component of position, Δx^1. Equation 29 therefore suggests that the **1**-component of momentum, p^1, ought to be related by two inertial observers, O and O′, in a way similar to the **1**-component of position, x^1. Now observers O and O′ connect coordinates x^0 (which is the new way of writing ct) and x^1 by the usual Lorentz transformation (see Equations 1a and 1b):

$$x'^1 = \gamma(V)\left(x^1 - \frac{V}{c}x^0\right) \tag{30a}$$

$$x'^0 = \gamma(V)\left(x^0 - \frac{V}{c}x^1\right). \tag{30b}$$

In these equations V is the velocity of O′ relative to O; the x^0 notation for ct emphasizes the equal footing of position and time. If we want to know the form of p^1 according to O′ we need merely find how Δx^1 transforms. From Equation 30a we have

$$\Delta x'^1 = \gamma(V)\left(\Delta x^1 - \frac{V}{c}\Delta x^0\right).$$

Dividing both sides by $\Delta\tau$ and multiplying by the mass m, gives

$$m\frac{\Delta x'^1}{\Delta \tau} = \gamma(V)\left(m\frac{\Delta x^1}{\Delta \tau} - \frac{V}{c}m\frac{\Delta x^0}{\Delta \tau}\right).$$

But the left-hand side of this equation is the **1**-component of linear momentum according to O′. We thus have

$$p'^1(\tau) = \gamma(V)\left(p^1(\tau) - \frac{V}{c}m\frac{\Delta x^0}{\Delta \tau}\right). \tag{31a}$$

Similarly, from Equation 30b one may show that

$$m\frac{\Delta x'^0}{\Delta \tau} = \gamma(V)\left(m\frac{\Delta x^0}{\Delta \tau} - \frac{V}{c}p^1(\tau)\right). \tag{31b}$$

Objectives 5 and 7   **SAQ 25**   Derive Equation 31b from Equation 30b.

By comparing Equations 30a and 30b to 31a and 31b it is clear that *the spacetime coordinates (x^0, x^1) transform in exactly the same way, from O* to O', as do the quantities $\left(m\dfrac{\Delta x^0}{\Delta\tau}, p^1\right)$. This suggests that the new

quantity, $m\dfrac{\Delta x^0}{\Delta\tau}$, may have physical significance.

In order to have a closer look at this new quantity, let us find an explicit form for it in some given inertial frame, O say. Since $\Delta x^0 = c\,\Delta t$, we have, remembering Equation 28,

$$m\frac{\Delta x^0}{\Delta\tau} = mc\frac{\Delta t}{\Delta\tau} = mc\,\gamma(v) \tag{32}$$

where v is the particle's speed according to O, and γ is the ubiquitous factor given, for instance, by Equation 27b.

To get more insight into the physical meaning of Equation 32, consider its right-hand side when the particle's speed v is small compared with c. To do this, we need merely expand $\gamma(v)$ in a series in powers of $(v/c)^2$. We find:

$$\gamma(v) \;=\; \frac{1}{\sqrt{1 - \left(\dfrac{v}{c}\right)^2}}$$

$$=\; 1 + \frac{1}{2}\left(\frac{v}{c}\right)^2 + \left(\text{terms like }\left(\frac{v}{c}\right)^4\text{ and higher}\right). \tag{33}$$

(Don't worry if such a *power series* is new to you and you are unable to derive it. In any application, you will be given the appropriate mathematical expansion. All you need to know is that if v/c is small, then terms containing high powers of v/c are so small that they can be neglected.)

We thus find that our new quantity becomes, for small values of v/c:

$$\left(m\frac{\Delta x^0}{\Delta\tau}\right) \approx mc + \frac{m}{2c}v^2 + \text{(higher order terms that can be neglected)}.$$

Suppose we multiply both sides by c. Then

$$c\left(m\frac{\Delta x^0}{\Delta\tau}\right) \approx mc^2 + \tfrac{1}{2}mv^2 + \text{(negligible higher order terms)}. \tag{34}$$

On the right-hand side of this equation we recognize a constant term, mc^2, plus the customary Newtonian kinetic energy, $\tfrac{1}{2}mv^2$. And then there are, of course, higher terms which are unimportant for slowly moving particles, $v \ll c$. This leads us to identify, in the context of special relativity,

energy of a free particle

$c(m\,\Delta x^0/\Delta\tau)$ *as the energy of a free particle of mass m and velocity v* (in the frame of O). This new form of energy now contains an extra, constant,

energy of particle at rest

term mc^2. **Thus the energy of the particle at rest is mc^2.** Even in Newtonian physics the energy of a particle can have an arbitrary constant added to it but now this new term takes on great significance. One very important reason that we include it in our new definition of energy is that (as we have shown with Equations 31a and 31b) it is the term $m\,\Delta x^0/\Delta\tau$, and not $(m\,\Delta x^0/\Delta\tau - mc)$, that transforms with the momentum p^1 in the same way as do spacetime coordinates (x^0, x^1).

Indeed, to emphasize this similarity between the pairs (x^0, x^1) and $\left(m\dfrac{\Delta x^0}{\Delta\tau}, p^1\right)$, we define this new quantity as $p^0(\tau)$ and refer to it as the

time component of momentum

'zero' or 'time' component of momentum:

$$\boxed{p^0(\tau) = m\frac{\Delta x^0}{\Delta\tau} = mc\,\gamma(v).} \tag{35}$$

energy of a free particle
(new expression)

> In special relativity, the energy of the particle, $c(m\,\Delta x^0/\Delta\tau)$, is then $cp^0(\tau)$.

The formal analogy between (x^0, x^1) and (p^0, p^1) is further strengthened by Theorem III, which parallels Theorem I for coordinates (x^0, x^1):

Theorem III

The quantity $(p^0(\tau))^2 - (p^1(\tau))^2$ is a Lorentz invariant, and its value is m^2c^2. That is, if (p^0, p^1) is the generalized momentum pair for O and (p'^0, p'^1) the corresponding pair for O$'$, we have:

$$(p^0(\tau))^2 - (p^1(\tau))^2 = m^2c^2 = (p'^0(\tau))^2 - (p'^1(\tau))^2. \tag{36}$$

The proof is easy. We apply the invariance of the interval ΔS connecting the two infinitesimally separated events shown on the particle's world-line in Figure 23 (this is a special case of Theorem I):

$$(\Delta S)^2 = c^2(\Delta t)^2 - (\Delta x^1)^2.$$

Multiply both sides by m^2c^2 and divide by $(\Delta S)^2 = (c\,\Delta\tau)^2$, to get

$$m^2c^2 = m^2c^2\left(\frac{\Delta t}{\Delta\tau}\right)^2 - m^2\left(\frac{\Delta x^1}{\Delta\tau}\right)^2.$$

Recognizing the expressions for p^1 from Equation 29 and for p^0 from Equation 35 (with $\Delta x^0 = c\,\Delta t$), we arrive at the left-hand side of Equation 36. But our proof holds for any inertial system, so we can say that the right-hand side of Equation 36 also holds. This completes the proof.

Remember that the mass m of a body is a Lorentz invariant property of a body, and in this respect is rather like the proper time of a clock fixed onto the body. The energy cp^0 and momentum p^1 will vary from inertial frame to inertial frame in just such a way that $(p^0(\tau))^2 - (p^1(\tau))^2$ is the same, i.e. m^2c^2, an invariant.

We summarize the close parallel between (x^0, x^1) and (p^0, p^1) in Figure 24.

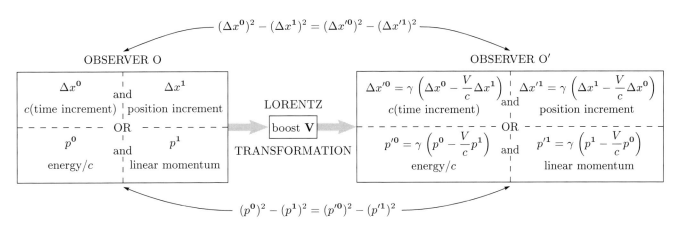

Figure 24 The Lorentz transformation for (time, position) is the same as for (energy, momentum).

Objective 6   SAQ 26   In a few sentences, describe the content of Figure 24.

We list briefly the main results of this Section, with some extra comments.

1 The linear momentum $p^1(\tau)$ of a particle along the **1**-axis of some inertial observer O, defined initially as $m\gamma(v)\Delta x^1/\Delta t$, may also be written as $m\,\Delta x^1/\Delta\tau$, where $x^1(\tau)$ is the x^1 coordinate as a function of proper time. Although we have written the expressions with $\Delta\tau$ etc., since this corresponds better to diagrams like Figure 17, one can always go to the limit where $\Delta \longrightarrow$ d everywhere.

2 The 'zero' or 'time' component of momentum of the particle is given by Equation 35:

$$p^0 = m\frac{\Delta x^0}{\Delta\tau} = mc\,\gamma(v).$$

3 The momentum components $(p^0(\tau),\, p^1(\tau))$ transform between any two inertial observers O and O$'$ in the usual way (Equations 31a and 31b). That is (see Equations 31 and 35),

$$p'^0(\tau) = \gamma(V)\left(p^0(\tau) - \frac{V}{c}p^1(\tau)\right) \tag{37a}$$

$$p'^1(\tau) = \gamma(V)\left(p^1(\tau) - \frac{V}{c}p^0(\tau)\right) \tag{37b}$$

where V is the velocity of O$'$ relative to O.

4 The energy of the free particle of speed v in some frame is defined as $cp^0(\tau) = mc^2\gamma(v)$. For small values of v/c it takes the form given in Equation 34. You should note that Equation 34 does *not* say that the energy equals mc^2 plus the kinetic energy plus a bit else. The kinetic energy of a freely moving body is *defined* as the total energy, cp^0, minus mc^2, the energy it would have at rest. That is to say, $\frac{1}{2}mv^2$ is not an exact expression for the kinetic energy when v/c becomes non-negligible. Although we refer to m as simply the mass of a body (the same in all frames) it is reasonable to refer to mc^2 as the *rest energy* of such a body since it is indeed the energy of a body with $v = 0$, i.e. $\gamma = 1$.

rest energy

Objectives 5 and 7   SAQ 27   Suppose we denote the energy of a particle by E. Show that for any observer O,

$$E^2 = (p^1(\tau))^2c^2 + m^2c^4.$$

Objective 7   SAQ 28   Show that p^1 becomes the usual Newtonian momentum when v/c is small. What is E when the particle is at rest relative to O?

6.4   Conservation of momentum (Theorem IV)

A cornerstone of Newtonian physics is the conservation of total linear momentum for a system of particles experiencing no *external* forces. This physical law is crucial in the analysis of the collision of two or more particles. Figure 25, for example, indicates a head-on collision (along the x-axis) between particle A of mass m and particle B of mass M.

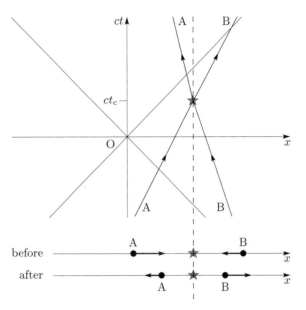

Figure 25   Head-on collision. The 'before' velocities apply when $ct < ct_c$, where t_c is the time of the collision. The 'after' velocities apply for $ct > ct_c$.

The collision, or interaction, between m and M is supposed to occur over a relatively localized region of space so that before and after the collision we can suppose that the particles move freely. Thus, although we may suppose that the actual process of collision is very complex, if Newtonian mechanics applies then we can connect the initial momenta, mu_i and MU_i, to the final momenta, mu_f and MU_f, as follows:

$$mu_i + MU_i = mu_f + MU_f \qquad \text{(in Newtonian mechanics)}.$$

The equation is written for the x- or x^1-component. The full vector equation would involve $\mathbf{u}_i, \mathbf{U}_i, \mathbf{u}_f$ and $\mathbf{U}_f$.

This is completely general and holds (in Newtonian mechanics) whatever the nature of the interaction process; the collision may be very nearly elastic as for two electrons or two billiard balls, or completely inelastic as for two balls of putty colliding and sticking together.

We now assert:

> In special relativity, the law of conservation of linear momentum carries over from the Newtonian conservation law in its most general form (that is, for any number of particles and for any type of collision *whether elastic or not*), provided the special relativistic definition for **p** (all four components) is used. Just as for the Newtonian law, the conservation law for linear momentum holds *in all inertial frames*.

The law of linear momentum conservation is one of the most far reaching laws of Nature. No deviation from it has ever been observed, on any scale from the submicroscopic to the cosmological, applying even to fundamental particles of velocities very close to c, i.e. with $\gamma(v) \gg 1$.

Momentum conservation has profound consequences: we shall now show that linear momentum conservation in collisions implies that the energy component p^0, is *also* conserved, whether or not the collision is elastic.

By asserting that, if linear momentum is conserved by a collision in one inertial frame, *then it is conserved in any other inertial frame*, we are simply acknowledging postulate SR1 which may be paraphrased as 'the laws of physics do not distinguish one inertial frame from any other'. Let us then apply this to the head-on collision of two particles, A and B, as in Figure 25. Let $(p_A^1)_i$ and $(p_A^1)_f$ denote the linear momentum of A according to observer O, before and after the collision respectively. Let the corresponding quantities according to observer O' be denoted by $(p_A'^1)_i$ and $(p_A'^1)_f$ in the usual way. Similarly we shall have $(p_B^1)_i$ and so forth for B. Then

$$(p_A^1 + p_B^1)_i = (p_A^1 + p_B^1)_f \qquad \text{(conservation of linear momentum according to O)} \qquad \text{(38a)}$$

$$(p_A'^1 + p_B'^1)_i = (p_A'^1 + p_B'^1)_f \qquad \text{(conservation of linear momentum according to O').} \qquad \text{(38b)}$$

In writing Equations 38a and 38b we have imposed postulate SR1, but we have not made use of the Lorentz transformation itself to connect the two observers O and O'. This we now do. We have from Equation 37b:

$$(p_A'^1)_i = \gamma(V)\left(p_A^1 - \frac{V}{c}p_A^0\right)_i \qquad \text{(39a)}$$

$$(p_A'^1)_f = \gamma(V)\left(p_A^1 - \frac{V}{c}p_A^0\right)_f \qquad \text{(39b)}$$

where $(p_A^0)_i$ and $(p_A^0)_f$ are the initial and final energy components of momentum for A.

Objectives 1 and 7   **SAQ 29**   Write down the corresponding initial and final equations for B.

We now use the two Equations 39a and 39b for particle A and the corresponding two equations from SAQ 29 for particle B to substitute for all the terms in Equation 38b.

We find:

$$\gamma(V)\left(p_A^1 - \frac{V}{c}p_A^0 + p_B^1 - \frac{V}{c}p_B^0\right)_i = \gamma(V)\left(p_A^1 - \frac{V}{c}p_A^0 + p_B^1 - \frac{V}{c}p_B^0\right)_f.$$

The factor $\gamma(V)$ may be cancelled from this equation. Then

$$(p_A^1 + p_B^1)_i \quad \text{and} \quad (p_A^1 + p_B^1)_f$$

may be subtracted from both sides, since these are equal according to Equation 38a (momentum conservation according to O).

We thus are left with

$$-\frac{V}{c}(p_A^0 + p_B^0)_i = -\frac{V}{c}(p_A^0 + p_B^0)_f$$

or,

$$(p_A^0 + p_B^0)_i = (p_A^0 + p_B^0)_f.$$

Given that energy $= cp^0$, and multiplying by c we get:

$$\boxed{(cp_A^0 + cp_B^0)_i = (cp_A^0 + cp_B^0)_f}$$ (conservation of energy according to O). (40a)

Clearly one can also apply the Lorentz transformation in the reverse direction, that is to say by expressing p_A^1 in terms of $p_A'^1$ and $p_A'^0$ and so forth, and multiplying by c, to obtain

$$\boxed{(cp_A'^0 + cp_B'^0)_i = (cp_A'^0 + cp_B'^0)_f}$$ (conservation of energy according to O'). (40b)

From these results you can see that by asserting that the conservation of linear momentum takes the same form in all inertial frames, as in Equations 38a and 38b, we have proved the following theorem:

Theorem IV

Energy is conserved and this conservation holds for all inertial observers, as in Equations 40a and 40b.

We can sum up what we have found in this section in one sentence: The fact that Lorentz transformations mix up the space-like and time-like components of (p^0, p^1) (in the same way that space and time (x^0, x^1) are mixed up by a Lorentz transformation), when combined with the fact that momentum p^1 is conserved, implies that *energy*, defined as cp^0, is also conserved.

6.5 $E = mc^2$ *(and all that)*

The proof in the preceding Section that the Lorentz invariance of momentum conservation implies the Lorentz invariance of energy conservation (Equations 40a and 40b) made fewer essential assumptions than you might have realized. In the first place, it was *not* assumed that either particle A or B had the *same* mass after the interaction as before. In the second place, particles A and B might interact to produce a single final particle of a new type; this could be accounted for by supposing either A or B has zero mass afterwards, leaving only one final particle with a new mass. Indeed, it would be possible, in so far as the momentum-energy equations are concerned, for A and B to create one or even several particles by their interaction; in the proof of Theorem IV we would just need to sum over the momenta and energies of all the initial and all the final particles.

Indeed, the creation of particles is a commonplace occurrence in experiments at high energies. In Figure 26 we present a picture showing particles created in a collision. A proton with energy several times the energy mc^2 which it has when at rest (i.e. a proton for which $\gamma \gg 1$) strikes another proton at rest in the target, and a whole shower of particles emerges. Their total energy is the same as the total energy before the collision, but much of the kinetic energy of the incoming particle has gone into the creation of new particles. We shall discuss the quantitative

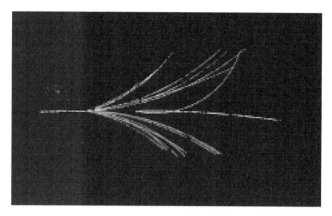

Figure 26   A shower of particles created when a very high energy proton collides with another in a 'bubble chamber'.

aspects of these reactions later. We would make just one point: whereas 'matter' can be created or destroyed, energy cannot. For a closed system the energy is the same for all time since p^0 is conserved in any fixed inertial frame even when it is the sum

$$p^0 = \sum_{i=1}^{N} p_i^0$$

over (a possibly variable!) number N of particles, labelled i.

First let us briefly summarize the essential tools required for seeing how energy conservation is applied in particular cases:

(i) The pair (p^0, p^1) for a particle are connected, in the usual way, by the Lorentz transformation, between any two inertial observers. For observers O and O' we can write

$$p'^0 = \gamma(V) \left(p^0 - \frac{V}{c} p^1 \right)$$
$$p'^1 = \gamma(V) \left(p^1 - \frac{V}{c} p^0 \right)$$

where V is the velocity of O' relative to O.

(ii) In any interaction, involving one or several particles, the total components (p^0 and p^1) are *separately* conserved even if the number of particles changes in the process under consideration.

(iii) For a particle of mass m and velocity $(v^1, v^2, v^3) = (u, 0, 0)$ relative to some inertial observer (along the observer's **1**-axis), the linear momentum (for this observer) is

$$p^1 = m \frac{\Delta x^1}{\Delta \tau} = mu\gamma(u)$$

and the energy is

$$E = cp^0 = mc \frac{\Delta x^0}{\Delta \tau} = mc^2 \gamma(u) = \frac{mc^2}{\sqrt{1 - \left(\frac{u}{c}\right)^2}}. \tag{41}$$

(iv) From Theorem III, the quantity $(p^0)^2 - (p^1)^2$ is a Lorentz invariant and equals $m^2 c^2$. For a single particle, m is the mass of the particle which has energy mc^2 in a frame in which the particle is at rest (i.e. in which $p^1 = 0$).

What we mean by the mass of a system of particles

> For a system of many particles, the *mass* m is again defined by considering the frame in which the total momentum p^1, summed over all the particles, is zero. The energy is just cp^0 in that frame, and the mass m is defined as energy over c^2, i.e. $mc^2 = cp^0$, where p^0 is the sum over all the particles.

Of course, if the number of particles changes during a collision (particles being created or destroyed) momentum conservation still applies. So, if the total momentum p^1 summed over all the particles is zero before the collision, it will be zero after the collision too; p^0 is also conserved in the collision. This expresses the fact that energy cp^0 is conserved, even when the number of particles contributing to p^0 changes during a collision. The examples below are intended to show how these ideas work out.

centre-of-momentum frame

(v) For a system of many particles in motion (gas molecules in a jar, or copper atoms in a piece of wire), it follows that the total mass is defined as follows: First, go to the frame (the *centre-of-momentum frame*) in which $\sum_i p_i^1 = 0$ (e.g. the jar of gas or piece of wire at rest); then, in this frame, $m = p^0/c$ where p^0 is the total over all particles. In this way, we shall be able to speak of the mass (of the jar of gas, say) increasing as it is heated, while the mass of each molecule, defined in the rest frame of that individual molecule, is invariant, as we have said. See Example 3 below.

Objectives 1 and 7   SAQ 30   Derive Equation 41. What does this equation suggest about the possibility of a material body travelling at the speed of light?

Just how these ideas work out is best seen in terms of examples. The first two examples concern collisions:

Example 1   The collision of two balls of putty

In this idealized (but possible) interaction, one putty ball, of mass m, strikes an identical ball lying at rest. Subsequently, they stick together to produce a single putty ball of some mass M. This is depicted in Figure 27. Throughout the following example, which is confined to the x-direction, we refer to the velocity using unbold characters u, v etc.

Figure 27   Collision in the laboratory frame; m has initial speed u and the conglomerate emerges with speed v.

Newtonian analysis

Let us first, for comparison, assume that Newtonian mechanics holds sway. Then we can say that $M = m + m$, for mass can be neither gained nor lost.

Furthermore, linear momentum is conserved, and therefore, in terms of the x-components,

$$mu = Mv = 2mv.$$

Thus we find that the outgoing (composite) particle moves away with speed $v = \frac{1}{2}u$.

Kinetic energy, however, is not conserved. To see this, subtract the final kinetic energy from the initial kinetic energy to find

$$\begin{aligned}
\text{loss of energy} &= \tfrac{1}{2}mu^2 - \tfrac{1}{2}Mv^2 \\
&= \tfrac{1}{2}mu^2 - \tfrac{1}{2}(2m)\left(\frac{u}{2}\right)^2 \\
&= \tfrac{1}{2}(\tfrac{1}{2}mu^2) \\
&= \tfrac{1}{2}(\text{initial energy}).
\end{aligned}$$

In Newtonian mechanics we say that kinetic energy is lost or converted into heat energy: the two putty balls generate heat on colliding. Recall that such a collision in which kinetic energy is lost is called an *inelastic collision*.

The collision according to special relativity

One must treat such a reaction according to special relativity if the speeds concerned are an appreciable fraction of c. This is almost always the case in fundamental particle physics; but even for slow speeds it is the more accurate analysis in principle. We express the conservation of linear momentum p^1 (using Equation 27) and of energy cp^0 (using Equation 41) as follows:

$$mu\gamma(u) = Mv\gamma(v) \quad \text{(conservation of momentum)} \tag{42}$$
$$mc^2\gamma(u) + mc^2 = Mc^2\gamma(v) \quad \text{(conservation of energy)} \tag{43}$$

(You should be able to write out these equations in terms of u and v, instead of $\gamma(u)$ or $\gamma(v)$; remember, too, that for a particle at rest, $\gamma = 1$).

In these two equations we regard the mass m, the incident velocity u, and of course c, as known or given parameters. It is algebraically complicated to find the unknown quantities M, the final composite mass, and v, the final velocity (strictly the x-component, but the other components are zero.) Before proceeding, we emphasize how important it is, as in the left side of Equation 43, to *retain* the energy mc^2 of the struck particle lying initially at rest, *for we can no longer assume that $M = m + m$.*

Objectives 1 and 7 **SAQ 31** Show that Equation 43 implies that $M = m + m$ in the non-relativistic limit where c greatly exceeds the velocities of the particles involved.

It turns out that the solution of Equations 42 and 43 leads to the following expressions for M and the final velocity, v:

$$\frac{v}{u} = \frac{\gamma(u)}{1 + \gamma(u)} \tag{44}$$

$$\frac{M}{m} = \sqrt{2}\,(1 + \gamma(u))^{1/2}. \tag{45}$$

Note that M *exceeds* $2m$ if u is not equal to zero, as can also be seen in Figure 28 where we have plotted Equations 44 and 45.

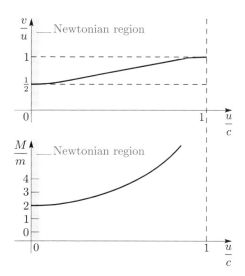

Figure 28   A plot of Equations 44 and 45. Remember, $\gamma(u)$ goes from $1 \longrightarrow \infty$ as u/c goes from $0 \longrightarrow 1$.

If we are thinking of balls of putty, the molecules comprising the final mass M have been excited to increase their kinetic energy (heat energy). But if the constituent atoms and molecules have increased their energy, then the composite putty ball has an increased relativistic energy, in any frame. In particular, in its rest frame, its energy is simply Mc^2. Therefore, since c is constant, its mass M must have been *increased* by the effect of heating. This explains why M always exceeds $2m$.

Study comment

Please read again the Study comment in the Introduction to Unit 7 with regard to the status of the following passage. Note that the subsequent optional passage makes use of the notation for centre-of-momentum introduced in this passage.

Non-assessable ▼
optional text

Let us now derive Equations 44 and 45 using a method based on the Lorentz invariance property, Equation 36.

Consider the collision from an inertial frame that is moving so that before the collision the total linear momentum is zero. This is called the 'centre-of-momentum system' (in the non-relativistic case it is the same as the centre-of-mass system.) We have denoted the frame depicted in Figure 27 by the subscript 'lab' and we shall denote the centre-of-momentum system by the subscript 'CM'. In the centre-of-momentum frame the collision occurs as in Figure 29.

We may think of the whole system as a composite particle; the effect of the interaction merely changes the character of this 'particle'. Both the lab and centre-of-momentum frames are inertial frames. From Figures 27 and 29 it is clear that the velocity of the centre-of-momentum system relative to the lab system is just v (the final velocity of the composite particle M, in the lab frame).

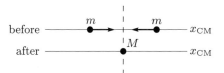

before m m x_{CM}

after M x_{CM}

Figure 29 Collision in the centre-of-momentum system.

A generally effective approach to such problems is to exploit the invariance of $(p^0)^2 - (p^1)^2$ between inertial frames. In our present case, p^0 and p^1 are taken to refer to the overall 'particle' consisting of two masses m beforehand and one mass M afterwards. We have:

$$(p^0)^2_{CM} - (p^1)^2_{CM} = (p^0)^2_{lab} - (p^1)^2_{lab}. \tag{46}$$

Such an equation applies both before and after the interaction. A second key point is that p^0 and p^1 are *separately* conserved by the process. After the interaction, in the centre-of-momentum system, we have a single mass M at rest. Thus from Equations 41 and 27, respectively, $(p^0)_{CM} = Mc$ and $(p^1)_{CM} = 0$. Similarly we may find expressions for our 'particle' in the lab system: $(p^0)_{lab} = mc\gamma(u) + mc$ and $(p^1)_{lab} = mu\gamma(u)$. Inserting these four quantities in Equation 46 we have:

$$(Mc)^2 = (mc\gamma(u) + mc)^2 - (mu\gamma(u))^2.$$

This equation can be solved fairly simply for the ratio M/m and gives the result previously quoted in Equation 45, namely

$$\frac{M}{m} = \sqrt{2}(1 + \gamma(u))^{1/2}.$$

An easy way to find the final velocity, v, of M in the lab frame is to make use of the Lorentz transformation properties of p^0 and p^1. In particular, we may use the Lorentz transformation on the **1**-component of momentum, with the primed frame taken as the centre-of-momentum frame and the unprimed frame taken as the lab frame. Since the centre-of-momentum system moves at velocity v relative to the lab system, we have:

$$(p^1)_{CM} = \gamma(v)\left((p^1)_{lab} - \frac{v}{c}(p^0)_{lab}\right).$$

The left-hand side is zero. Using the previously established values for $(p^1)_{lab}$ and $(p^0)_{lab}$ gives:

$$0 = \gamma(v)\left\{mu\gamma(u) - \frac{v}{c}mc(1 + \gamma(u))\right\}.$$

End of optional text ▲   This leads immediately to the previous result, Equation 44.

Example 2 Threshold energy

In fundamental particle physics it is common to fire one particle against another in order to create fundamental particles of new and interesting types by the resulting interaction. As a simple example, consider firing a proton against another proton. If the incident proton has sufficient energy, it is possible to produce *four* particles: three protons, p, and one 'antiproton', $\overline{\text{p}}$. (Other particles can be created, too, if there is sufficient energy. All the different possible outcomes occur with probabilities determined by quantum theory, but it is perfectly legitimate to consider, as here, the energy–momentum balance for one particular possible reaction.) The required minimum incident energy is called the 'threshold energy' for the reaction $\text{p} + \text{p} \rightarrow \text{p} + \text{p} + \text{p} + \overline{\text{p}}$. The antiproton has the same mass as a proton but the opposite charge, so charge is conserved in this process. If one is constructing a particle accelerator to bring about a process such as this, one cannot avoid special relativity. (In 1955, Chamberlain and Segré created the first antiprotons at Berkeley, leading to a Nobel prize.)

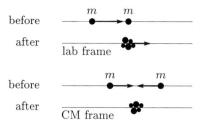

Figure 30   The creation of new particles, at threshold energy.

threshold energy

Consider, then, the experiment in Figure 30 where we imagine firing a mass m against another mass m to produce several particles, possibly of various different masses.

Let us suppose that the incident particle has been given just sufficient energy, E, to be able to produce the new particles (E is the *threshold energy* in the lab frame for the reaction). Then the final group of particles will have no final motion relative to each other but will move off as a bunch as in Figure 30. Let us also suppose that we know the net sum of all the final masses. Denote this sum by $\sum m_k$, indicating a summation over all the various final masses m_k for $k = 1, 2, 3$, etc.

Study comment

Remember what the Study comment in the introduction said about the status of the following passage.

Non-assessable ▼
optional text

As in the previous example, we make use of the Lorentz invariance relationship. Considering the overall system of incident plus target particle as a particle with energy $cp^{\mathbf{0}}$ and momentum $p^{\mathbf{1}}$, we have,

$$(p^{\mathbf{0}})^2_{\text{lab}} - (p^{\mathbf{1}})^2_{\text{lab}} = (p^{\mathbf{0}})^2_{\text{CM}} - (p^{\mathbf{1}})^2_{\text{CM}}.$$

If the energy of the incident particle in the lab is E, then $c(p^{\mathbf{0}})_{\text{lab}} = mc^2 + E$. In the centre-of-momentum system the momentum is zero, i.e. $(p^{\mathbf{1}})_{\text{CM}} = 0$ and, since the newly created particles do not move apart (we are operating just at threshold) the final centre-of-momentum energy is $c(p^{\mathbf{0}})_{\text{CM}} = (\sum m_k)c^2$. We need only find an expression for $(p^{\mathbf{1}})_{\text{lab}}$ in terms of E and m. Two points to remember are: (i) E, the energy of the incident particle, is *not* the kinetic energy of the incident particle ... that kinetic energy is, of course, $E - mc^2$, and, (ii) it is never helpful to speak of 'mass being converted to energy' or vice versa, though *material particles* can be created using energy.

Objectives 1, 7 & 8   SAQ 32   Show that for the composite particle (whole system),

$$(p^{\mathbf{1}})^2_{\text{lab}} = \frac{(E^2 - m^2c^4)}{c^2}.$$

We have now evaluated all terms in the invariance relationship. It becomes

$$\frac{1}{c^2}(mc^2 + E)^2 - \frac{1}{c^2}(E^2 - m^2c^4) = (\sum_k m_k)^2 c^2$$

easily giving the following simple equation for threshold energy E:

$$2mc^2(E + mc^2) = (\sum_k m_k c^2)^2. \tag{47}$$

Objectives 6, 7 & 8   SAQ 33   Describe the meaning of Equation 47. What would happen if the incident energy were greater than the value of E implied by this equation?

Objectives 6, 7 & 8 **SAQ 34** Consider the reaction $p + p \to p + p + p + \bar{p}$, where we are firing a proton against a stationary proton in the laboratory. The mass m of an antiproton, $\bar{p}$, is equal to that of a proton, p. Find the threshold energy, E. Why is this greater than the sum of the energies associated with the mass of the incident proton plus the energy of the newly created proton and antiproton (namely $3mc^2$)?

End of optional text ▲

Example 3 The mass of heat

As a next example showing the expression $E = mc^2$ in action, consider a block of matter which is irradiated uniformly from all sides so that it absorbs heat energy E_{rad}. How much does its mass increase? (The radiation is uniform from all sides because radiation does convey momentum, and we don't want to confuse the issue by having the block of matter recoil; we assume that its total momentum is zero in our observation frame both before and after the heating process.)

We hinted in point (iv) just before Example 1 what the key to this question is. First let us assume that all the energy goes into increasing the kinetic energy of all the atoms and electrons in the block (this is not strictly true, but does not affect the answer; our argument would apply exactly to an idealized gas of particles in a container, for example, in which the gas molecules collide but otherwise have no interacting forces). Now, since we are looking at the block from a frame in which its total momentum $p^1 = 0$ (i.e. $\sum_i p_i^1 = 0$ with the sum over all atoms, i) we have

$$m^2 c^2 = (p^0)^2 - (p^1)^2 = (p^0)^2 = (\sum_i p_i^0)^2 \qquad (48)$$

from which we can say that the mass of the block, m is just

$$m = \frac{p^0}{c} = \frac{cp^0}{c^2} = \frac{E}{c^2}.$$

Moreover, since all of the atoms will be moving just that bit faster after the absorption of the heat energy, their γ factors will have increased a little, and each individual p_i^0 will have increased a little bit. The net effect is that as the energy E increases by E_{rad}, the mass of the block increases by E_{rad}/c^2.

Example 4 Mass of a hydrogen atom

As a final example, we note that when a proton captures an electron to make a hydrogen atom in its 'ground state', light is emitted corresponding to a release of energy $= 13.6\,\text{eV}$. (An electronvolt, eV, is 1.6×10^{-19} joule.) We say that the 'binding energy' of a hydrogen atom is $E_{\text{binding}} = 13.6\,\text{eV}$, and that is also the minimum energy that is required to ionize an H atom. The hydrogen atom has a mass which is less (in the centre-of-momentum frame) than the sum of the individual masses of the electron and proton by E_{binding}/c^2. This is quite a small amount. However E_{binding} for nuclei is typically 10^6 times higher than for atoms, and this is what accounts for the fact that the energy released in nuclear reactions is greater, by a similar ratio, than that released in chemical reactions. Notice that if the hydrogen atom is in a closed container when it is formed, the total mass of the *system*, including the container, is fixed, since the 13.6 eV of energy radiated as the atom forms will be absorbed by the walls of the container, which will warm up and increase in mass by an amount equal to the loss of mass by the electron and proton.

You will notice one ongoing theme in the above examples: the total energy, E, of a *closed* system in its centre-of-momentum frame, and its mass, E/c^2, both remain constant. Mass converts from one form to another, as does energy. But we prefer not to speak of mass turning into energy. Rather it is *matter* that can appear or disappear, with corresponding changes in the amount of such forms of energy as kinetic energy, potential energy, etc. So, for example, when a positron and electron annihilate to form two photons these are subsequently absorbed to warm up the surroundings. But the mass is just transferred to the surroundings as the γ factors for the atoms in the surroundings increase a little in accord with Equation 48. The following SAQs are important, and we hope they will throw some further light on these questions, so do read the solutions even if you cannot do them at first.

Objective 8   SAQ 35   (a) Using the above figure for the binding energy, and also relevant data on the back covers of your Blocks, calculate how much less the mass of a hydrogen atom is than the sum of the mass of a proton and an electron. Express this both in kilograms, and as a percentage of the total mass (the last answer need only be correct to two or three significant figures).

(b) Given that it requires 2.2 MeV (million electronvolts) of energy to split a deuteron (nucleus of heavy hydrogen) into a proton and neutron, by how much is the mass of a deuteron less than the mass of a proton and neutron?

Objective 8   SAQ 36   Consider a stationary container (sealed bottle, perhaps) containing N identical molecules of gas each of mass m. The gas is at non-zero temperature (i.e. the molecules are in random motion). Is the mass of the gas (i) Nm; (ii) greater than Nm; or (iii) less than Nm? Explain your answer fully (you can ignore any interaction between molecules).

7   The ultimate speed

No particle has ever been observed to travel at a speed greater than the speed of light, and the impossibility of 'superluminal' speeds is now an established part of modern physics. There are three lines of argument for this which can be based on what we have seen in this Block:

In the first place we showed (Section 3.3 of Unit 6) that in passing from one frame to another in a sequence of uniformly moving frames, it is impossible to achieve a speed, between the first and last frames of the sequence, exceeding c.

In the second place you saw, from SAQ 30, and Equation 41, that a particle with any non-zero mass would need to acquire *infinite* energy in order to attain speed c.

Finally, in Section 6 of this Unit, you saw that the linear momentum in an inertial frame, defined by

$$p = m \frac{u}{\sqrt{1 - \left(\dfrac{u}{c}\right)^2}} = mu\gamma(u)$$

carries great physical significance. While it is true that this reduces, for u/c small, to the Newtonian definition, mu, it is also true that it increases without bound for values of $(u/c)^2$ approaching unity. Since we lack space to consider the relativistic generalization of Newton's second law, we cannot *prove* that an infinite force is required to accelerate a material particle to the speed c. Nevertheless, it is the case that this new property of momentum ensures that no material particle shall be accelerated to a speed exceeding c; for in special relativity, force (appropriately defined) still equals the rate of change of linear momentum, and one cannot conceive of a force of infinite magnitude in a physical Universe.

Of course, a particle with mass zero, such as a photon (the light quantum) not only *can* but *must* travel at the speed of light in a vacuum. It is also the case that information cannot travel faster than light; signals propagating outside the light cone could lead to effects preceding causes. But what about a spot of light on the clouds caused by a searchlight? Is there any limit to how fast (in principle, at least) it could move across the clouds if you could twist the searchlight rapidly enough about its axis? No. In fact, the spot has no mass so there are no energy or momentum problems, and its motion cannot convey information across the cloud; in fact relativity imposes no limit on the speed of abstract entities of the kind 'point where light beam intersects the cloud'.

8 Relativity and electromagnetism

Our entry route into relativity was through electromagnetism; the very title of Einstein's original paper makes relativity a topic within electromagnetism. We cannot leave special relativity without making a few remarks concerning this subject.

From what we have seen in Units 4 and 5, it appears that a pure electric field in one inertial frame becomes a superposition of magnetic and electric fields in another inertial field: $\mathbf{E}$ and $\mathbf{B}$ become mixed up. But having seen first space and time become mixed up with each other, and then momentum and energy, that should not be so shocking. Clearly, special relativity must provide a systematic way of transforming $\mathbf{E}$ and $\mathbf{B}$ into one another when we change inertial frames. Once we have this, we would hope to be able to recast Maxwell's equations into a form which makes it easy to show that they are indeed of the same form when we transform from one inertial frame to another, i.e. that they are Lorentz invariant. It is far from obvious from an inspection of Maxwell's equations that they can be written in such a way. But they can. Indeed, it is possible to construct the whole of electromagnetic theory from the two requirements:

(i) that the force between two charges at rest obeys Coulomb's law and,

(ii) that the laws be of a form which is the same in each inertial frame, i.e. obey SR1.

Magnetism from this new perspective is a relativistic phenomenon.

These ideas put a new complexion on the question of why the speed of light appears everywhere in the theory of relativity. One can say that the basic point is that the structure of spacetime is determined by the nature of the invariant interval:

$$(\Delta S)^2 = c^2(\Delta t)^2 - ((\Delta x^1)^2 + (\Delta x^2)^2 + (\Delta x^3)^2)$$

where c is a constant characterizing spacetime and having dimensions of speed. Then, when one constructs a theory which is like Coulomb's law in the rest frame of two charges and also Lorentz invariant, the quantity c in the expression for ΔS pops up in the theory and emerges as the speed of waves predicted by the theory.

The key to constructing a manifestly Lorentz invariant electromagnetism is through the introduction of potentials; this is beyond the scope of this Course, but some flavour of how the theory goes can be given as follows. Recall from Unit 4 that the source terms of Maxwell's equations involved a current density vector $\mathbf{j}$ and a charge density scalar ρ. It turns out that not only do $\mathbf{E}$ and $\mathbf{B}$ get mixed up like position and time when one looks at a system from a different inertial frame, but so do $\mathbf{j}$ and ρ: indeed the foursome $(c\rho, j^1, j^2, j^3)$ transforms in exactly the same way that is now familiar for the time–position and energy–momentum 'four-vectors'.

Finally, the complete relativistic electromagnetism abolishes the *instantaneous* action at a distance that Coulomb's law seems to indicate. All electromagnetic effects are transmitted at speed c (in a vacuum — the speed is less in any other medium). Indeed, c is the maximum speed with which *any* information can travel. Modern relativistic quantum electrodynamics (QED) pictures electromagnetic waves and forces as carried by massless elementary particles called *photons*. (Recall that the arguments of Section 7 do not forbid a massless particle from travelling at speed c.) It is the photon picture that finally disposes of the old conundrum — if there is no ether, what carries light in a vacuum?

9 Unit Summary

1 Spacetime is a four-dimensional mathematical structure consisting of the set of all events, analogous to the customary three-dimensional Euclidean space which is the set of all points in space. The distance between two points in three-dimensional space is invariant (unchanged) by any rotation. One can define an invariant 'distance', in four-dimensional spacetime, between any two events $\mathscr{E}_a$ and $\mathscr{E}_b$. Its square is

$$(S_{ab})^2 = (ct_b - ct_a)^2 - (x_b^1 - x_a^1)^2 - (x_b^2 - x_a^2)^2 - (x_b^3 - x_a^3)^2 \qquad (6)$$

and it is *invariant* with respect to the Lorentz transformation (Theorem I).

2 Relative to any particular event $\mathscr{E}_a$, the spacetime diagram divides itself up into distinct regions. There are two regions such that an event $\mathscr{E}_b$ is related to $\mathscr{E}_a$ by an interval, S_{ab}, that is *time-like*, i.e. such that $(S_{ab})^2$ is positive. One of these two regions contains events lying in the *absolute future* with respect to $\mathscr{E}_a$, and the other contains events lying in the *absolute past* with respect to $\mathscr{E}_a$. In other regions, however, the event $\mathscr{E}_b$ is related to $\mathscr{E}_a$ by a *space-like* interval, i.e. $(S_{ab})^2$ is negative. Such events cannot be *causally related*, as different inertial observers may disagree on whether $\mathscr{E}_b$ lies in the future or the past with respect to $\mathscr{E}_a$. The time-like and space-like regions are separated by the *light-cone* surface, on which any pair of events is separated by an interval that has zero value, i.e. $(S_{ab})^2 = 0$.

3 By use of the concept of an *instantaneous rest frame*, special relativity can be applied to problems involving accelerating frames of reference. Observations made from an accelerating frame at a particular instant are the same as those made by an inertial observer from an instantaneous rest frame travelling with the same velocity at that instant.

4 A particle's world-line can be regarded as made up of individual events separated by arbitrary *small* intervals, ΔS. The *proper time* between two such events is defined by $\Delta\tau = \Delta S/c$. By summing all such small intervals between two arbitrary events $\mathscr{E}_c$ and $\mathscr{E}_d$, we get the total proper time τ_{cd} between such events. (The proper time τ_{cd} is the time elapsed on a clock moving with the particle.) The proper time can be used to label, in an invariant way, the progress of a particle along its world-line. For instance, in two-dimensional spacetime, an observer O would be able to write the particle's coordinates as $x^0(\tau)$ and $x^1(\tau)$, and any other observer O′ would write $x'^0(\tau)$ and $x'^1(\tau)$. The proper time is always equal to or less than the time recorded by an inertial observer observing the same two events $\mathscr{E}_c$ and $\mathscr{E}_d$ (Theorem II).

It follows from this that in the so-called twin paradox, the twin who performs the out-and-back journey must come back younger than the one who stays at home. This conclusion is confirmed by observations of the average lifetime of muons coasting at high speed around a storage ring.

5 An equation was derived for the *relativistic Doppler shift* appropriate to arbitrary motion in three-dimensional space (Equation 25).

6 A relativistic version for the equation for *momentum* was obtained:

$$\mathbf{p} = m\gamma(v)\,\mathbf{v}.$$

It was noted that this reduces to the Newtonian form ($\mathbf{p} = m\mathbf{v}$) in the limit $v \ll c$.

7 The relativistic form of the **1**-component of momentum can be written as:

$$p^1 = m \left(\frac{\Delta x^1}{\Delta \tau} \right). \tag{29}$$

Because both m and τ are Lorentz invariant, it would seem that p^1 should transform like the position x^1. Examination of the resulting transformation equations led us to the quantity $mc\,\Delta x^0/\Delta \tau$, or $m\gamma c^2$, and this became identified with the special relativistic form of the expression for *energy* (Equation 34). It was noted that in the limit $v \ll c$, the relativistic expression for energy approximated to $mc^2 + \frac{1}{2}mv^2$, where the second term is the Newtonian expression for kinetic energy and the first term is a constant.

Writing the energy E as $cp^0(\tau)$, we noted that the quantity $\{(p^0(\tau))^2 - (p^1(\tau))^2\}$ is a Lorentz invariant with a value of $m^2 c^2$ (Theorem III).

8 The law of *conservation of momentum* was restated in its special relativistic version.

The law of *conservation of energy* was obtained by requiring the momentum to transform according to the Lorentz transformations, and conservation of linear momentum to take the same form *in all inertial systems* (Theorem IV).

9 It was pointed out that the proof of Theorem IV actually still holds even if one allows for the creation and annihilation of the interacting particles. It appears then, that matter can be converted into energy and vice versa. That this happens in Nature is well attested.

The law of momentum and energy conservation in special relativity was illustrated with four examples.

In the first example two balls of putty collide and stick together. This is manifestly not an elastic collision, but the laws of momentum and energy conservation still apply. The final, composite, particle has a mass *larger* than the sum of the masses of the initial particles. This is due to the conversion of some of the initial motional, or kinetic, energy into energy associated with heat by the process of frictional, inelastic, heating. The mass of a heated object is greater than the mass of the same object cold.

The second example carried out a calculation of the *threshold energy* for the creation of several new particles by the collision of two initial particles. The threshold energy is the *minimum* incident energy required for the process to proceed.

The third example discussed the increase in mass occuring when a body is heated.

The fourth example discussed the binding energy of atoms and nuclei. Changes in binding energy can result in energy being released — rather more for nuclear than for chemical transformations.

10 Special relativity extends beyond our treatment. Einstein showed how to modify Newton's law to describe the motion of charged particles under electromagnetic forces. Solutions to this equation show that material particles never move faster than c. This is not too surprising in view of the fact that the energy of a particle, γmc^2, and its linear momentum, $\gamma m\mathbf{v}$, become infinite for values of v approaching c.

Band 8 of AC2 comments on this Unit.

9.1   Key concepts for later Units

This has been a long Unit with many new ideas. Some of these could easily be overlooked in the excitement of seeing a discussion of $E = mc^2$. As promised, we now mention some of the concepts which will be important later. The following are particularly important in Units 9–12: *proper time*; *instantaneous rest frame*; *Theorem II* and *transverse Doppler effect*. In Unit 12, on black holes, a useful concept will be that of *light cone*.

Self-assessment questions — answers and comments

SAQ 1 From Equation 1a we have:

$$(x'_b - x'_a) = \gamma(V)\{(x_b - x_a) - V(t_b - t_a)\}.$$

From Equation 1b, multiplying by c, we have:

$$(ct'_b - ct'_a) = \gamma(V)\left\{(ct_b - ct_a) - \frac{V}{c}(x_b - x_a)\right\}.$$

Squaring both equations above and subtracting the first from the second gives:

$$(ct'_b - ct'_a)^2 - (x'_b - x'_a)^2 =$$

$$\gamma^2(V)\left\{(ct_b - ct_a)^2 + \frac{V^2}{c^2}(x_b - x_a)^2\right\}$$

$$- \gamma^2(V)\left\{(x_b - x_a)^2 + V^2(t_b - t_a)^2\right\}$$

(the cross products cancel).

The right-hand side becomes:

$$\gamma^2(V)\left\{\left(1 - \frac{V^2}{c^2}\right)(ct_b - ct_a)^2 - \left(1 - \frac{V^2}{c^2}\right)(x_b - x_a)^2\right\}.$$

Since $\gamma^2(V) = \dfrac{1}{1 - \left(\dfrac{V}{c}\right)^2}$, this equals

$$(ct_b - ct_a)^2 - (x_b - x_a)^2$$

which was to be proved.

SAQ 2 $(S_{ab})^2$ is invariant under arbitrary spatial rotations since

$$\{(x^1_b - x^1_a)^2 + (x^2_b - x^2_a)^2 + (x^3_b - x^3_a)^2\}$$

is the *distance squared* between two points in three-dimensional space, and is unchanged by rotations (see Figure 3, for example). Thus any expression for $(S_{ab})^2$ must draw no distinction between directions in space and must satisfy the condition of isotropy. Similarly, the distance between two points is unaffected by translations of the spatial axes. Thus any expression for $(S_{ab})^2$ has to satisfy the condition of spatial homogeneity.

SAQ 3 In the new notation (using Equations 1a and 1b or, better, your memory!) we have:

$$x'^1 = \gamma(V)\left(x^1 - \frac{V}{c}x^0\right)$$

$$x'^0 = \gamma(V)\left(x^0 - \frac{V}{c}x^1\right).$$

From Equation 6, for the general case (Figure 6) we have

$$(S_{ab})^2 = (x^0_b - x^0_a)^2 - (x^1_b - x^1_a)^2 - (x^2_b - x^2_a)^2 - (x^3_b - x^3_a)^2.$$

SAQ 4 In SAQ 1 you showed that $(x^0_b - x^0_a)^2 - (x^1_b - x^1_a)^2$ is invariant, namely

$$(x^0_b - x^0_a)^2 - (x^1_b - x^1_a)^2 = (x'^0_b - x'^0_a)^2 - (x'^1_b - x'^1_a)^2.$$

By Theorem I, $(S_{ab})^2$ is also invariant, namely

$$(x^0_b - x^0_a)^2 - (x^1_b - x^1_a)^2 - (x^2_b - x^2_a)^2 - (x^3_b - x^3_a)^2$$
$$= (x'^0_b - x'^0_a)^2 - (x'^1_b - x'^1_a)^2 - (x'^2_b - x'^2_a)^2 - (x'^3_b - x'^3_a)^2.$$

We can subtract the two equations expressing the invariance to get

$$(x'^2_b - x'^2_a)^2 + (x'^3_b - x'^3_a)^2 = (x^2_b - x^2_a)^2 + (x^3_b - x^3_a)^2.$$

Now this must be true for any values of x_b, x_a, etc. It is clear that the simplest relationships satisfying this equation are $x'^2_b = x^2_b$ and $x'^3_b = x^3_b$, etc.

SAQ 5 Figure 31 is a sketch of the world-line.

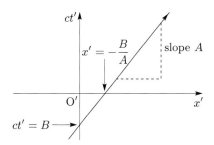

Figure 31 For SAQ 5.

SAQ 6 From Equation 1b, $t' = 0$ implies that $t = Vx/c^2$, or $ct = Vx/c$. From Equation 1a, $x' = 0$ implies $x = Vt$, or $ct = cx/V$. These are shown in Figure 32. If V/c were small, the lines representing the ct'- and x'-axes would 'open out' to correspond (very nearly) to the ct- and x-axes as in Figure 32(b).

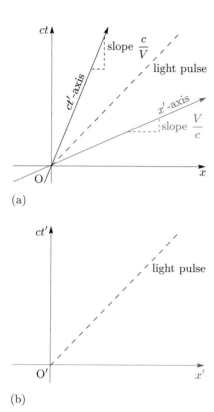

(a)

(b)

Figure 32 For SAQ 6.

SAQ 7 All events in region III of Figure 11 can be characterized by an event such as $\mathscr{E}_d$ in Figure 10, where it is clear that the x part of the coordinate rectangle of $\mathscr{E}_d$ is longer than the ct part. Thus, for all events like $\mathscr{E}_d$ we have $(x)^2 > (ct)^2$ or $(S)^2 < 0$, and such points are space-like with respect to $(0,0)$. Similarly, all events in region IV like $\mathscr{E}_c$ are space-like.

SAQ 8 The areas that are space-like or time-like with respect to $\mathscr{E}_a$ are as shown in Figure 33.

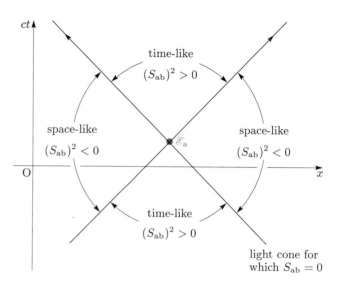

Figure 33 For SAQ 8.

SAQ 9 Any event on the light cone of O is characterized by $(ct)^2 - (x)^2 = 0$, so that $(S)^2 = 0$. However, S is a Lorentz invariant, so $(ct')^2 - (x')^2 = 0$. In other words, events on the light cone of O also lie somewhere on the light cone of O'.

SAQ 10 Events in the absolute future and absolute past for $\mathscr{E}_a$ would be recorded in the areas indicated in Figure 34.

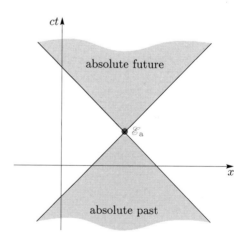

Figure 34 For SAQ 10.

SAQ 11 Equation 1b applied to event $\mathscr{E}_c$ of Figure 10 gives

$$ct'_c = \gamma(V)\left(ct_c - \frac{V}{c}x_c\right).$$

Since (for $\mathscr{E}_c$) $x_c > ct_c$, we can find some (positive) value for V such that $V/c < 1$ but $t' < 0$. In particular, any value of V/c exceeding ct_c/x_c will do the job.

SAQ 12 Since the particle must move at a speed less than c its world-line must lie within the light cone from its starting point, here event $\mathscr{E}_a$. This means that *all* events *on the particle's world-line* are time-like with respect to *any* earlier event, and hence to event $\mathscr{E}_a$.

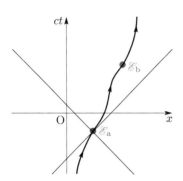

Figure 35 For SAQ 12.

SAQ 13 From Equations 14 and 16 we have $c^2(\Delta t')^2 = c^2(\Delta t)^2 - (\Delta x)^2$. Equation 17 follows by dividing both sides by c^2, factoring out $(\Delta t)^2$ from the right-hand side and taking the square root of both sides. Equation 18 follows from Equation 17 by replacing $\Delta t'$ by $\Delta S/c$.

SAQ 14 Using $x^0 = ct$ gives

$$(\Delta S)^2 = (\Delta x^0)^2 - \left\{ (\Delta x^1)^2 + (\Delta x^2)^2 + (\Delta x^3)^2 \right\}.$$

SAQ 15 The interval squared, $(\Delta S)^2 = (c\Delta t)^2 - (\Delta x)^2$, is a Lorentz invariant. Particles are assumed to travel at speeds less than c. Thus it is possible to classify events on a particle's world-line as having an order, and this order will be the same for all inertial observers. For any small section ΔS of the world-line, we can imagine an instantaneous rest frame moving along with the particle. In this frame, O', the particle lies at rest so $\Delta S = c\,\Delta t'$. Since ΔS is invariant, $c\,\Delta t' = \sqrt{(c\,\Delta t)^2 - (\Delta x)^2}$. The term $\Delta t'$ is defined as the proper time increment. Thus proper time is just the elapsed time in the particle's (instantaneous) rest frame, and is denoted $\Delta \tau$.

SAQ 16 The proper time elapsed between $\mathscr{E}_c$ and $\mathscr{E}_d$ is the time elapsed on a clock moving with the particle. Since this clock always *advances* as the particle moves along, all observers must agree as to the *sense* of motion. But, since in any frame $u(t) \leqslant c$ (as we saw in Section 3.3 of Unit 5, or from the fact that τ would become imaginary from Equation 22 and also for dynamical reasons to be revealed in Section 6), the world-line of the particle in Figure 18 must stay within the light cone, and thus be in the absolute future when τ increases.

SAQ 17 Using the values for τ and $\gamma(u)$ given immediately before the question, from the equation the dilated decay time is

$$T = (2.196\,66) \times (29.33) \times 10^{-6}\,\text{s} = 64.426 \times 10^{-6}\,\text{s}.$$

This is very close to the measured value of $(64.368 \pm 0.029)\,\text{s}$, amply verifying that the twin 'paradox' is no paradox, but rather an intrinsic property of spacetime.

SAQ 18 The proper time increment, $\Delta\tau$, between two adjacent points on a particle's world-line is reduced by a factor $\sqrt{1 - (u/c)^2}$ from the inertial observer's recorded time, Δt, between the same two points.

SAQ 19 When the source moves along the line of sight towards the observer we have $\theta = 0$ (see Figure 19). Then Equation 25 becomes

$$\Delta t = \Delta \tau \frac{\left(1 - \dfrac{u}{c} \right)}{\sqrt{1 - \left(\dfrac{u}{c} \right)^2}} = \Delta \tau \sqrt{\frac{1 - \dfrac{u}{c}}{1 + \dfrac{u}{c}}}.$$

where Δt is the time between received pulses. The right-hand side of this equation is of the same form as the right-hand side of Equation 41 in Unit 6 (except that there we called the proper time T instead of $\Delta\tau$).

SAQ 20 The analysis parallels that in Section 2.2 of Unit 6. Consider the interval of time between two crests of the wave. If the frequency is f_0, this time interval is $\Delta\tau = 1/f_0$. But the radiation also moves at speed c with respect to the receiving data taker. If his received frequency is f, then for him, the time between the arrival of two crests is $1/f$, where

$$\frac{1}{f} = \frac{1}{f_0} \frac{\left(1 - \dfrac{u}{c}\cos\theta \right)}{\sqrt{1 - \left(\dfrac{u}{c} \right)^2}}.$$

SAQ 21 The momentum of the *ball* is not conserved when it hits the wall since its velocity and momentum will be more or less reversed. But the momentum of the *total system* is conserved. The wall will recoil and carry off the missing momentum. The recoil will probably be imperceptible because of the wall's large mass which might even be considered to be that of the Earth if the wall is firmly attached to the Earth. When applying conservation laws, whether of charge, energy, momentum or whatever, it is important to make sure that you know what the relevant complete isolated system is.

SAQ 22 Frame O$'$ moves at velocity $\mathbf{W}$ relative to O along x^1. In frame O the transverse velocity of A$'$ after collision is Ω. In frame O$'$ the transverse velocity of A$'$ after collision is u, and the longitudinal component of its velocity is zero (see Figures 21 and 22). From Equation 31 of Unit 6 we have the connecting equation for the transverse **2**-component of the particle's velocity between frames O$'$ and O:

$$U'^2 = \frac{U^2}{\gamma(W)\left(1 - \dfrac{WU^1}{c^2}\right)}.$$

Here U^1 is the **1**-component of the particle's velocity in frame O. To make use of the fact that the **1**-component of the velocity of A$'$ after collision is zero we can make use of the corresponding equation, but going from frame O$'$ to frame O:

$$U^2 = \frac{U'^2}{\gamma(W)\left(1 + \dfrac{WU'^1}{c^2}\right)}.$$

Since $U'^1 = 0$, $U^2 = \Omega$ and $U'^2 = u$ in the present example, we have $\Omega = u/\gamma(W)$, as required.

SAQ 23 There are two physical assumptions involved in this thought-experiment based on elastic scattering. First, we assumed a form $mf(v)\mathbf{v}$ for linear momentum. Second, and most important perhaps, we assumed that linear momentum is conserved by an elastic collision.

We further go on to assume that conclusions drawn from a consideration of elastic collisions are generally valid.

SAQ 24 Of course, just as in Newton's mechanics, an observer may consider all of the particle's variables, p, x, etc., as functions of time t. This can be awkward since t is different for different inertial observers. However, the time t and proper time τ can be related functionally as follows. Since $\Delta S = c\,\Delta\tau$ and $(\Delta S)^2 = c^2(\Delta t)^2 - (\Delta x)^2$ from Equation 13,

$$(c\,\Delta\tau)^2 = c^2(\Delta t)^2 - (\Delta x)^2.$$

In this equation, Δt and Δx are the increments of time and position (according to O) between two adjacent events on the particle's world-line. Thus, dividing by $c^2(\Delta t)^2$,

$$\frac{\Delta\tau}{\Delta t} = \sqrt{1 - \left(\frac{\Delta x}{\Delta t}\right)^2 \frac{1}{c^2}} = \sqrt{1 - \left(\frac{u(t)}{c}\right)^2}$$

where $u(t)$ is the particle's velocity (according to O). Taking the limit $\Delta t \longrightarrow 0$ gives an equation relating t and τ (assuming that $u(t)$ is known, from measurement say):

$$\frac{d\tau}{dt} = \sqrt{1 - \left(\frac{u(t)}{c}\right)^2} = \frac{1}{\gamma(u(t))}.$$

SAQ 25 Using the substitutions $t' = x'^0/c$ and $t = x^0/c$, Equation 30b relates the coordinates x^0 and x^1 to x'^0. It clearly applies to coordinate increments as well. Thus

$$\Delta x'^0 = \gamma(V)\left(\Delta x^0 - \frac{V}{c}\Delta x^1\right).$$

Divide both sides by $\Delta\tau$ and multiply by the particle's mass, m. Substituting the linear momentum, $p^1(\tau) = m\,\Delta x^1/\Delta t$, then gives the required result, Equation 31b.

SAQ 26 Figure 24 makes several points. First, the pair (p^0, p^1) are connected by any two inertial observers in the *same* way as are the pair (x^0, x^1), namely the Lorentz transformation. Second, the Lorentz transformation for the case of two such observers moving relatively along their mutual **1**-axes is given in the Figure (and you should have memorized it by this time). Third, the interval $(\Delta x^0)^2 - (\Delta x^1)^2$ is a Lorentz invariant, where $(\Delta x^0, \Delta x^1)$ is the difference in coordinates between *any* two events. Fourth, the quantity $(p^0)^2 - (p^1)^2$ is also a Lorentz invariant, where p^0 and p^1 are the time and space components of the generalized momentum of a particle, $cp^0(\tau)$ is the relativistic energy and $p^1(\tau)$ is the linear momentum.

SAQ 27 The result follows from Equation 36. For any observer O we have

$$(p^0(\tau))^2 = m^2c^2 + (p^1(\tau))^2.$$

But the energy (according to O) is $E = cp^0$. Thus

$$E^2 = m^2c^4 + c^2(p^1(\tau))^2.$$

SAQ 28 From Equation 27a, $\mathbf{p} = m\gamma(v)\,\mathbf{v}$. But from Equation 33, $\gamma(v) \approx 1$ for small v. This gives $\mathbf{p} \longrightarrow m\mathbf{v}$ as $v \longrightarrow 0$, so $p^1 \longrightarrow mv^1$. For a particle at rest in O, $E = mc^2$.

SAQ 29 $(p_B'^1)_i = \gamma(V)\left(p_B^1 - \dfrac{V}{c}p_B^0\right)_i$

$(p_B'^1)_f = \gamma(V)\left(p_B^1 - \dfrac{V}{c}p_B^0\right)_f.$

SAQ 30 The energy is defined as $E = cp^0$. From Equation 35, $p^0 = m\,\Delta x^0/\Delta\tau = mc\gamma(v)$. Putting these three items together and substituting for $\gamma(v)$ (or $\gamma(u)$ as we are calling it in the present case) from Equation 27b gives

$$E = cp^0 = mc\frac{\Delta x^0}{\Delta\tau} = mc^2\gamma(u) = \frac{mc^2}{\sqrt{1 - \left(\dfrac{u}{c}\right)^2}}.$$

Since $E \longrightarrow \infty$ as $u \longrightarrow c$, infinite energy would need to be given to a material, i.e. $m \neq 0$, particle for it to reach the speed of light. Of course, photons, the quanta of light, do travel, in a vacuum, only at speed c, but they are massless particles, so in Equation 41 E can be finite if both the denominator and numerator are zero.

SAQ 31 For v/c so small it can be ignored, Equation 43 reduces to

$$mc^2 + mc^2 = Mc^2 \quad \text{or} \quad M = 2m.$$

SAQ 32 In the laboratory, only the incident particle moves. If its linear momentum is called p, we can use the equation from SAQ 27:

$$E^2 = m^2c^4 + p^2c^2.$$

Here E is the energy of the incident particle in the lab system and m is its mass. Now consider the overall system as a particle. In the lab frame this particle has linear momentum $(p^1)_{\text{lab}} = p$, since the struck particle lies at rest. Thus

$$(p^1)^2_{\text{lab}} = \frac{(E^2 - m^2c^4)}{c^2}.$$

SAQ 33 Equation 47 describes the required incident energy that a particle of mass m must have, in colliding with a stationary particle (also of mass m), to produce a number of masses, m_k, moving as a group (i.e bearing no excess kinetic motion of separation). E is called the threshold energy for the reaction

$$m + m \longrightarrow (m_1 + m_2 + m_3 + \cdots).$$

If the incident particle had (in the lab) an energy exceeding E, the particles produced might move apart, or other additional particles might be produced, or some combination of these two possibilities might occur.

SAQ 34 From Equation 47

$$2mc^2(E + mc^2) = \left(\sum_k m_k c^2\right)^2 = (4mc^2)^2 = 16m^2c^4$$

giving

$$E = 7mc^2.$$

This exceeds the value $3mc^2$ because of the law of conservation of linear momentum. Since this law must be satisfied in the lab system, for example, a certain amount (here $4mc^2$) of the incident particle's energy is 'tied up', to provide the linear momentum of the final products. The final products of the interaction must move off to the right (in the lab) so as to conserve linear momentum, and their resulting kinetic energy is this 'tied up' energy.

SAQ 35 (a) Since $1\,\text{eV}$ is $1.6 \times 10^{-19}\,\text{J}$, and $c = 2.998 \times 10^8$ m s^{-1}, the mass difference is

$$\frac{13.6 \times 1.6 \times 10^{-19}}{(2.998 \times 10^8)^2}\,\text{kg} = 2.42 \times 10^{-35}\,\text{kg}.$$

Since the electron is much less massive than a proton, we can get a good approximation for the percentage by using the proton mass as the H atom's mass. Since the mass of a proton is 1.67×10^{-27} kg, the mass corresponding to the binding energy of a H atom is about 1.45×10^{-6} per cent of the mass of the atom.

(b) One need only calculate the mass equivalent of $2.2\,\text{MeV}$, which is

$$\frac{2.2 \times 10^6 \times 1.6 \times 10^{-19}}{(2.998 \times 10^8)^2}\,\text{kg} = 3.9 \times 10^{-30}\,\text{kg}.$$

We didn't ask it, but this is some 0.115 per cent of the deuteron mass. The fact that binding energies are so much larger for nuclei than for atoms or molecules is the reason that much less energy is released from chemical processes (e.g. TNT, coal burning) than nuclear processes (e.g. uranium fission).

SAQ 36 The molecules will all move with different velocities but you can suppose that since the bottle of gas is stationary, the total (vector) momentum of the molecules is zero. The frame of the bottle is then the centre-of-momentum frame of the gas so that the total energy cp^0 is just the sum of terms like $m\gamma(v_i)c^2$ where v_i is the speed of the ith molecule. Now we know that, always, $\gamma(v_i) > 1$ so the average of $\gamma(v_i)$ over all molecules is $\bar{\gamma} > 1$. The total energy of the molecules will be $Nmc^2\bar{\gamma} > Nmc^2$. So the total mass is $Nm\bar{\gamma}$ and this is greater than Nm. Again, you can think of the excess as the mass of heat energy.

Unit 8 Consolidation and revision I

Prepared by the Course Team

Contents

1 Introduction

You have now completed your work on the first seven units. At this point in the course, a short period has been set aside for revision and consolidation of the material covered so far. This provides an opportunity for you to strengthen and deepen your understanding, and enable you to gain more confidence. Unit 8 contains nothing essentially new; its purpose is simply to assist in structuring the revision process. You will almost certainly find that there is more material here than you can use at the moment, but you will probably find it useful to return to it in the period between the submission of the last assignment and the examination.

The primary source of revision material is, of course, the units themselves. Perhaps the most effective use of this unit is as a means of identifying those parts of Units 1–7 where rereading may be necessary. We suggest that you tackle this work in four stages, as detailed below.

1 First of all, make sure you can answer all the SAQs in Units 1–7, or at least understand the answers provided.

2 As a quick check on your grasp of the basic concepts, their definitions and meaning, a set of 40 comprehension exercises, based on each of the first two blocks, is given in Sections 2 and 4. Brief, rapid-access answers to all these exercises may be found in the relational glossaries for Blocks 1 and 2.

3 The principles and techniques developed in the units are more fully explored in the consolidation exercises in Sections 3 and 5. They fall into the three broad categories listed below, according to purpose:

(i) to reinforce understanding of a basic concept;

(ii) to provide practice in the use of a technique;

(iii) to present an alternative approach to a theoretical argument, e.g. by joining the chain of logic at a slightly different point.

These categories are not mutually exclusive, and a particular exercise may belong to more than one of them. Broadly speaking, the exercises are specific to a particular unit. They have been designed in this way to help you identify the basic principles on which the solution depends, but this is not intended to encourage you to compartmentalize your knowledge. The units are convenient subdivisions, not self-contained modules. Ideally, the course should be seen as a seamless whole.

Use of asterisk ***** *in questions*

The exercises are not all of the same length and difficulty. While some are relatively short and straightforward like SAQs, others will require more thought, approaching TMA question difficulty. The longer exercises are subdivided into short sections, a few of which are marked with an asterisk. *The asterisked sections are a little more demanding, and it is quite safe to ignore them, at least at the first reading.* As with the SAQs in the units, a detailed solution is provided for each exercise, and these are intended for flexible use. Your method of working will obviously depend on personal preference, and the amount of useful time you have available. You may enjoy working completely through a long exercise before consulting the given solution. Alternatively, you may prefer to compare your results with the worked solution after each short section. In some cases, you may find it useful just to read through the solution, although this is unlikely to benefit you much unless you give the question some careful thought beforehand. It must be emphasized that these exercises are not, in any sense, a test and that both they and the solutions are intended to help you to master the material in the units.

4 You will be required to write an essay in Part III of the examination, which is worth a quite significant fraction of the total marks, and past experience indicates that some students perform poorly on this part of the examination as a result of inadequate technique. While we cannot give you a crash-course on essay writing in this unit, we are able to provide some guidance. Almost 80% of the marks for the essay are awarded for *content*, i.e. the relevant points mentioned in answer to the question. The remainder of the marks are set aside for other aspects such as organization of the material and clarity.

Section 7 consists of a number of essay exercises, and we recommend that you attempt one or two of these under examination conditions. By comparing the points you have made with those given in the notes provided, you will be able to estimate what fraction of the 80% of the credit, referred to above, you would actually receive. Clearly, you will have to be your own judge of whether you have presented your points in a clear, intelligible manner. The essay topics in Section 7 are not necessarily specific to a particular unit, and may involve material from both Blocks 1 and 2. Essay topics ranging across different units, and even blocks, are not unusual in the examination.

Spacetime diagrams tutorial   Section 6 does not consist of exercises, but instead is a tutorial based on a series of spacetime diagrams, each with an associated commentary. This material was included in the latter years of the predecessor course and was considered quite useful by many students. Although placed here as Section 6, you may wish to read this section *before* doing the consolidation exercises in Sections 5.3 and 5.4, some of which explicitly involve spacetime diagrams.

2 Comprehension exercises for Block 1

Complete answers to these exercises may be found in the Glossary for Block 1 by looking up the word or phrase displayed in capital letters.

1 Define the ACCELERATION, **a**, of a particle, and state the conditions under which two inertial observers will agree about its magnitude.

2 Write down an expression for the ANGULAR MOMENTUM, **J**, of an isolated system of two particles. Under what conditions does the law of CONSERVATION OF ANGULAR MOMENTUM apply to such a system?

3 What is meant by ANGULAR SPEED?

4 Give a definition of each of the following principles:

(i) HOMOGENEITY OF TIME

(ii) HOMOGENEITY OF SPACE

(iii) ISOTROPY OF SPACE.

5 Match each of the following conservation laws with the symmetry principle(s) in Exercise 4 of which it is a manifestation:

(i) CONSERVATION OF ANGULAR MOMENTUM

(ii) CONSERVATION OF ENERGY

(iii) CONSERVATION OF MOMENTUM.

6 Write down the components of the DISPLACEMENT vector from position P to position Q in terms of the components of the position vectors of P and Q. How is the DISTANCE between P and Q related to the components of the displacement vector?

7 Define an ELLIPSE, and explain what is meant by the FOCI and the ECCENTRICITY, ε. What are the limiting values of ε?

8 Distinguish between the terms KINEMATICS, DYNAMICS and MECHANICS.

9 Define the term FICTITIOUS FORCE, and give an alternative name by which such forces are known. How does a fictitious force depend on the inertial mass of the body on which it acts?

10 Explain what is meant by the FORM INVARIANCE of a physical law under a transformation relating inertial frames of reference. Illustrate the concept using Newton's law of universal gravitation as an example.

11 Write down the GALILEAN TRANSFORMATION between two identically calibrated reference frames.

12 Distinguish between a FRAME OF REFERENCE and a COORDINATE SYSTEM.

13 State GALILEO'S FIRST LAW of motion (G1).

14 State GALILEO'S SECOND LAW of motion (G2).

15 Describe how the GRAVITATIONAL MASS, μ, of a particle may be defined.

16 Describe an operational procedure for establishing an INERTIAL FRAME OF REFERENCE using two free particles.

17 Complete the sentence below with **four** alternative endings. 'An INERTIAL FRAME OF REFERENCE is the only type of frame of reference in which ...'.

18 Give an operational definition of the INERTIAL MASS, m, of a particle. In Newtonian mechanics, how is the inertial mass related to the gravitational mass?

19 Explain what is meant by an INVARIANT under a particular transformation between inertial frames of reference.

20 State KEPLER'S FIRST LAW of planetary motion (K1).

21 State KEPLER'S SECOND LAW of planetary motion (K2).

22 State KEPLER'S THIRD LAW of planetary motion (K3).

23 KEPLER'S THREE LAWS imply **two** facts about the acceleration vector of a planet. What are they?

24 Define the MAGNITUDE $|\mathbf{a}|$ of a vector $\mathbf{a}$, in terms of its components.

25 Write down an expression for the MOMENTUM, P, of an isolated system of two particles. Under what conditions does the law of CONSERVATION OF MOMENTUM apply?

26 State NEWTON'S FIRST LAW of motion (N1).

27 State NEWTON'S SECOND LAW of motion (N2).

28 State NEWTON'S THIRD LAW of motion (N3).

29 Which of NEWTON'S FIRST, SECOND and THIRD LAWS of motion can be regarded as:

(i) a definition of force?

(ii) an expression of the homogeneity and isotropy of space?

(iii) defining an inertial frame of reference?

30 Write down an expression for the force of gravity exerted on particle 1 by particle 2 according to NEWTON'S LAW OF UNIVERSAL GRAVITATION.

31 Explain what is meant by a NON-INERTIAL FRAME OF REFERENCE.

32 Complete the following sentence. 'The POTENTIAL ENERGY FUNCTION, U, of an isolated system of particles depends only on ...'.

33 Explain what the PRINCIPLE OF RELATIVITY in Newtonian mechanics means in operational terms.

34 Define the SCALAR PRODUCT, $\mathbf{a} \cdot \mathbf{b}$, of two vectors $\mathbf{a}$ and $\mathbf{b}$ in terms of their components. Give a simple expression for $\mathbf{a} \cdot \mathbf{b}$ in terms of the magnitudes of $\mathbf{a}$ and $\mathbf{b}$ and the angle between them.

35 Explain what is meant by a STATIC TRANSFORMATION between two frames of reference.

36 State the law of SUPERPOSITION OF FORCES.

37 Explain the terms UNIFORM ANGULAR BOOST and UNIFORM LINEAR BOOST. Which of the two is described, in Newtonian mechanics, by the Galilean transformation?

38 Explain the terms UNIFORMLY ACCELERATED MOTION and UNIFORM CIRCULAR MOTION of a particle. In which, if either, is the speed of the particle constant in time?

39 Define the VECTOR PRODUCT, $\mathbf{a} \times \mathbf{b}$, of two vectors $\mathbf{a}$ and $\mathbf{b}$. Give a simple formula for its magnitude in terms of the magnitudes of $\mathbf{a}$ and $\mathbf{b}$ and the angle between them. What is its direction?

40 Explain what is meant by the WORLD-LINE of a particle.

3   Consolidation exercises for Block 1

Introductory notes

A basic concept in S357 is that of a 'reference frame' to which the position of a particle may be referred and relative to which its motion may be studied. This idea is introduced in Section 3.2 of Unit 1. The position vector of a point can be specified either through its Cartesian components, (x^1, x^2, x^3), or the spherical polar coordinates r, θ and ϕ. Exercise 1 deals with the relationship between these two systems of coordinates.

Vectors play a central role in Blocks 1 and 2, and several of the exercises are devoted to this important topic. The examples cover position, displacement, velocity and acceleration vectors and are intended to demonstrate the power of the vector method in solving problems involving more than one spatial dimension. If you do not have extensive experience in the use of vectors, you may need to give these exercises especially close attention.

The defining characteristic of a vector is the manner in which its components change when the coordinate axes undergo a static rotation. This and other properties of vectors are explored in Exercises 9 and 10. Other quantities, known as *scalars*, have the property of being *invariant* under rotations, i.e. unchanged by any rotation of the coordinate axes. These include the magnitudes or lengths of vectors and the angle between two vectors. The scalar product of two vectors $\mathbf{a}$ and $\mathbf{b}$ is defined by

$$\mathbf{a} \cdot \mathbf{b} = a^1 b^1 + a^2 b^2 + a^3 b^3.$$

As its name suggests, this quantity is invariant under rotations, and is equal to the product of the magnitudes of $\mathbf{a}$ and $\mathbf{b}$ and the cosine of the angle between them. It therefore provides a means of calculating the angle between two given vectors, as explained in Section 6.3 Unit 1. Exercise 11 is concerned with invariants in a slightly wider context.

Two vectors, $\mathbf{a}$ and $\mathbf{b}$, can also be combined to form the quantity

$$\mathbf{a} \times \mathbf{b} = (a^2 b^3 - a^3 b^2, a^3 b^1 - a^1 b^3, a^1 b^2 - a^2 b^1),$$

the so-called vector product. This is a vector in so far as its behaviour under rotations of the coordinate axes is concerned. The vector product $\mathbf{a} \times \mathbf{b}$ can be thought of as a vector perpendicular to both $\mathbf{a}$ and $\mathbf{b}$, pointing in the direction indicated by the middle finger of a right hand, in which the thumb represents $\mathbf{a}$ and the index finger represents $\mathbf{b}$. It is clear from the definition that

$$\mathbf{a} \times \mathbf{b} = -\mathbf{b} \times \mathbf{a},$$

and that the vector product of parallel vectors is zero.

The study of dynamics begins in Unit 2 with the analysis of Kepler's laws of planetary motion. Section 2.2 in that unit involves some of the most difficult mathematics in the course. Exercise 12 provides a slight amplification of the material on the geometry of ellipses given in the text, while Exercise 13 concerns a physical system obeying laws rather similar to those of Kepler, though somewhat simpler. This section ends with a couple of simple examples on the dynamics of circular motion.

Unit 3 opens with a discussion of Galilean relativity, and Exercise 16 explores this topic. The remaining exercises in the section concern the application of Newton's laws, including N3, to systems of particles and the derivation of the conservation laws of momentum, angular momentum and energy.

3.1   Unit 1

Exercise 1   (i) The position of a point A is given in spherical polar coordinates by $(r, \theta, \phi) = (5, 112°, 37°)$. What are the Cartesian coordinates of A? (ii) The Cartesian coordinates of another point B are $(x^1, x^2, x^3) = (2, -3, 7)$. Give the position of B in spherical polar coordinates.

Exercise 2   A ship sails 80 nautical miles due north from A to B, then changes course and sails 100 nautical miles north-east to C. From there it sails 50 nautical miles due west to D. How far is D from A, and in which direction? If the speed is a steady 22 knots, what is the average velocity of the ship from A to D? [1 knot = 1 nautical mile per hour; assume the ship is not too near the North Pole.]

Exercise 3   The position vectors of three points A, B and C are given by $(1, 1, 1)$, $(1, -1, -1)$, and $(-1, 1, -1)$ respectively. Write down the displacement vectors $\overrightarrow{AB}$, $\overrightarrow{BC}$ and $\overrightarrow{CA}$, and hence calculate the lengths of all the sides and the internal angles of the triangle ABC.

Exercise 4   (i) Show that the position of a point P, on the surface of the Earth can be represented by a vector $(R \sin \theta \cos \phi, R \sin \theta \sin \phi, R \cos \theta)$ where R is the radius of the Earth (assumed spherical), θ is the *co-latitude* of P (which means that the latitude is $90° - \theta$), and ϕ is related to the longitude of P. The Cartesian axes are defined relative to the centre of the Earth as origin, with the x^1- and x^2-axes crossing the Equator at longitude zero and $90°$ E respectively, while x^3 lies along the polar axis of the Earth. Write down the values of θ and ϕ for London (latitude $51°$ N, longitude zero) and Montevideo (latitude $35°$ S, longitude $56°$ W).

(ii) How can the vectors for two cities be used to calculate their distance apart, by the great circle route? Calculate the distance apart of London and Montevideo in nautical miles. [A nautical mile at the surface subtends an angle of 1 minute of arc at the centre of the Earth. You may find the trigonometric relation $\cos \alpha \cos \beta + \sin \alpha \sin \beta = \cos(\alpha - \beta)$ useful.]

Exercise 5   (i) A geostationary satellite, S, is located over the Equator at longitude $60°$ E. Assuming that the satellite is a distance $6.626R$ from the centre O of the Earth (where R is the radius of the Earth), write down its position vector, $\overrightarrow{OS}$, in Cartesian form using the geocentric coordinate system of Exercise 4.

(ii) Write down the position vector, $\overrightarrow{OL}$, of London, L, and hence obtain the displacement vector $\overrightarrow{LS} = \overrightarrow{OS} - \overrightarrow{OL}$ from London to the satellite. Calculate the distance from London to the satellite.

(iii) Calculate the angle between $\overrightarrow{LS}$ and $\overrightarrow{OL}$, and show that a satellite dish in London should be pointed at an elevation of approximately $10°$ above the horizon to receive signals from the satellite.

* **Exercise 6**   The crew of a submarine, which is just below the surface of the sea, spot an enemy warship 2.5 nautical miles due north of them steaming at a steady 22 knots on a north-easterly course. They immediately fire a torpedo, which hits the warship. If the torpedo travels at 55 knots, in what direction was it fired, and how long did it take to reach the target?

Exercise 7   (i) If the position of a particle is given by

$$\mathbf{x} = (-3 \sin \omega t, 4 \sin \omega t, 5 \cos \omega t)$$

where ω is a constant, write down the vectors for its velocity, $\mathbf{v}$, and acceleration, $\mathbf{a}$.

(ii) Show that $\mathbf{x} \times \mathbf{a} = \mathbf{0}$ and that $\mathbf{x} \times \mathbf{v}$ does not depend on the time t. Explain the connection between these two results.

(iii) Evaluate the scalar products $\mathbf{x} \cdot \mathbf{x}$, $\mathbf{x} \cdot \mathbf{v}$ and $\mathbf{v} \cdot \mathbf{v}$ and use them to show that the particle is moving in a circle with constant speed.

* (iv) Think of the particle as fixed to the rim of a rotating wheel. In what direction is the axle of the wheel pointing? Is the wheel rotating in a clockwise or anticlockwise sense?

Exercise 8   (i) The position vector of a particle is given by

$$\mathbf{x} = (R(\omega t - \sin \omega t), R(1 - \cos \omega t), 0)$$

where R and ω are constants. Write down its velocity vector, $\mathbf{v}$, and its acceleration vector, $\mathbf{a}$.

(ii) Find the times and positions at which the particle is instantaneously at rest, and show that $\mathbf{a} = (0, R\omega^2, 0)$ at these times.

* (iii) Sketch the path of the particle from $t = 0$ to $t = 4\pi/\omega$.

Exercise 9   (i) Show that a rotation through π radians about the **3**-axis, followed by a reflection in the **1,2**-plane is equivalent to simultaneous reversal of all three axes, i.e. $(x^1, x^2, x^3) \longrightarrow (-x^1, -x^2, -x^3)$.

* (ii) Show that a positive rotation through $\pi/2$ radians about the **3**-axis followed by a rotation through π radians about the *new* **1**-axis has the effect

$$(x^1, x^2, x^3) \longrightarrow (x^2, x^1, -x^3).$$

[*Note*: The *sense* of a rotation about an axis is given by the right-hand rule — when the right thumb is directed along the axis in the positive sense, the fingers curl in the direction of a positive rotation.]

Exercise 10   (i) In a Cartesian reference frame S, the position and velocity of a particle at a given moment are $\mathbf{x} = (1, -2, 4)$ and $\mathbf{v} = (-5, 2, 0)$ respectively. Write down the values these vectors would have if referred to another Cartesian frame S′ obtained from S by a static transformation T. Consider separately the cases where T is

(a) a rotation through π about the **1**-axis;

(b) a positive rotation through $\pi/2$ about the **2**-axis;

(c) a reflection in the **1,2**-plane;

(d) a translation of the origin by $(3, 3, 1)$.

(ii) Show that $\mathbf{x} \cdot \mathbf{v}$ is invariant under (a)–(c).

Exercise 11   The key below shows an assortment of five quantities expressed in terms of the vectors for position, $\mathbf{x}$, velocity, $\mathbf{v}$, and acceleration, $\mathbf{a}$, of a particle. Write down the key letters for those which are *invariant* under the following static transformations of the coordinates:

(i) a recalibration of the time scale from seconds to nanoseconds;

(ii) a recalibration of the distance scale from metres to micrometres;

(iii) a translation of the origin of coordinates;

(iv) a rotation of the axes about the origin.

KEY for Exercise 11

A $|\mathbf{a}|/|\mathbf{v}|^2$

B $|\mathbf{a}|^2/(\mathbf{x} \cdot \mathbf{v})$

C $\mathbf{a}/(\mathbf{x} \cdot \mathbf{v})$

D $\mathbf{v} \times \mathbf{a}$

E $\mathbf{x} \cdot (\mathbf{v} \times \mathbf{a})$

3.2   Unit 2

Exercise 12   (i) The Cartesian equation for the ellipse in Figure 1, referred to the centre as origin is

$$\left(\frac{x^1}{r_1}\right)^2 + \left(\frac{x^2}{r_2}\right)^2 = 1.$$

Prove this equation by expressing x^1 and x^2 for the point P in terms of r_1, r_2 and θ (cf. SAQ 1 Unit 2), and using the identity $\cos^2\theta + \sin^2\theta = 1$.

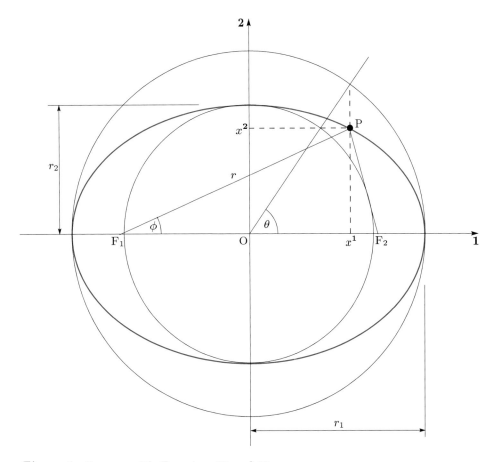

Figure 1   For use with Exercises 12 and 13.

* (ii) A simple method for drawing an ellipse using two drawing pins and a loop of string is based on the equation $|F_1P| + |F_2P| = 2r_1$, where F_1 is the left-hand focus and F_2 the right-hand focus of the ellipse. Prove this equation.
[*Hint*: Use Pythagoras' theorem to show that $|F_1P| = r_1(1 + \varepsilon \cos \theta)$ and obtain a similar result for $|F_2P|$ (cf. SAQ 2 Unit 2).]

* (iii) Show that if the position of the point P is given by polar coordinates $r = |FP|$ and $\phi = \angle OFP$ (i.e. the angle OFP), where F is either of the two foci, then $r(1 - \varepsilon \cos \phi) = r_1(1 - \varepsilon^2)$. This is the polar equation of the ellipse, referred to a focus as origin.

Exercise 13   A physicist studies the motions of a number of particles which move in planar orbits about a fixed point, O. From his observations, he deduces the following three laws.

Law 1 Each orbit is an ellipse *centred* at O.

Law 2 The radius vector joining the particle to O sweeps out equal areas in equal times.

Law 3 The orbital period T is a constant, *independent* of the size of the orbit.

(i) Referring to Figure 1, let the particle be at P. Using Law 1, write down the position vector, **x**, of the particle in terms of the angle θ. Obtain the velocity, **v**, by differentiation.

(ii) Use **x** and **v** to calculate the rate at which area is swept out by the line joining the particle to O. Hence show that Law 2 implies that

$$\frac{d\theta}{dt} = \text{constant},$$

and obtain its value in terms of T. [*Hint*: You may assume that the area of the ellipse is $\pi r_1 r_2$.]

(iii) Obtain an expression for the magnitude, $|\mathbf{a}|$, of the acceleration of a particle in terms of T and $|\mathbf{x}|$. What is the direction of the acceleration vector?

Exercise 14   (i) A small satellite orbits at a height h above the surface of a smooth, spherical, airless planet of radius R. If $h \ll R$ and the acceleration due to gravity at the surface of the planet is g, show that the orbital period, T, of the satellite is given by

$$T = 2\pi \sqrt{R/g}.$$

Taking the values $R = 6.37 \times 10^6$ m and $g = 9.8 \, \text{m s}^{-2}$, calculate the orbital period of a *hypothetical* satellite orbiting at a very low height above the Earth's surface. (A real satellite would burn up in the atmosphere or crash into the Andes!)

(ii) Use your result, together with Kepler's third law, to determine the orbital radius of a geostationary satellite.

[*Hint*: Work out the acceleration which the satellite must have towards the centre of the planet by virtue of its orbital motion. The force sustaining this acceleration is provided by the gravitational attraction between the satellite and the planet. Assume the period of the orbit is exactly 24 hours.]

Exercise 15 An astronaut in training sits in a pod at the end of a rotor-arm which rotates in a horizontal plane at constant angular speed. If the horizontal force on the astronaut is equal to nine times his own weight, and the length of the rotor-arm is 5 m, at what angular speed is it rotating?

3.3 Unit 3

Exercise 16 (i) An observer, O, using an inertial reference frame, S, studies a collision between two *identical* billiard balls A and B, which move so that their centres are confined to the **1**,**2**-plane. He notes that at $t = t_1$, before the collision, their positions and velocities in his frame are

$$\mathbf{x}_A = (-X, 0, 0) \quad \mathbf{v}_A = (V, 0, 0)$$
$$\mathbf{x}_B = (0, -X, 0) \quad \mathbf{v}_B = (0, V, 0).$$

A second inertial observer O′, whose reference frame S′ differs from S by interchange of the **1**- and **2**-axes, also studies the collision. Show that, provided O′ interchanges the names of the colliding balls, her description of the initial configuration of the system is identical to that of O.

(ii) At $t = t_2$, after the collision, O observes that

$$\mathbf{x}_A = (a, b, 0) \quad \mathbf{v}_A = (u, v, 0).$$

According to the principle of relativity, what result must he obtain for $\mathbf{x}_B$ and $\mathbf{v}_B$?

(iii) Assuming that no energy is lost in the collision, use momentum and energy conservation to obtain the values of u and v.

* (iv) Is it then possible to calculate either a or b from angular momentum conservation?

Exercise 17 (i) An isolated system consists of two particles of masses m_1 and m_2. The force exerted on particle 1 by particle 2 is

$$\mathbf{F}_{12} = f(|\mathbf{x}|) \, \frac{\mathbf{x}}{|\mathbf{x}|},$$

where $\mathbf{x} = \mathbf{x}_1 - \mathbf{x}_2$. Use a simple symmetry argument to obtain $\mathbf{F}_{21}$, the force exerted on particle 2 by particle 1, and confirm that Newton's third law (N3) is obeyed.

(ii) According to Newton's second law (N2), the motion of the particles is governed by the equations of motion

$$m_1 \frac{d^2\mathbf{x}_1}{dt^2} = f(|\mathbf{x}|) \, \frac{\mathbf{x}}{|\mathbf{x}|} \quad \text{and} \quad m_2 \frac{d^2\mathbf{x}_2}{dt^2} = -f(|\mathbf{x}|) \, \frac{\mathbf{x}}{|\mathbf{x}|}.$$

By adding these two equations, show that the centre of mass of the system, defined by

$$\mathbf{X} = \frac{m_1\mathbf{x}_1 + m_2\mathbf{x}_2}{M},$$

where $M = m_1 + m_2$, moves with constant velocity.

(iii) Show that, if the centre of mass of the system is chosen as the origin of coordinates, then

$$\mathbf{x}_1 = \frac{m_2}{M}\mathbf{x} \quad \text{and} \quad \mathbf{x}_2 = -\frac{m_1}{M}\mathbf{x},$$

and *both* equations of motion become

$$\mu \frac{d^2\mathbf{x}}{dt^2} = f(|\mathbf{x}|) \, \frac{\mathbf{x}}{|\mathbf{x}|},$$

where μ is the *reduced mass* of the system, given by

$$\frac{1}{\mu} = \frac{1}{m_1} + \frac{1}{m_2}.$$

* (iv) Consider a binary star system where the force acting is gravity, for which

$$f(|\mathbf{x}|) = -\frac{Gm_1m_2}{|\mathbf{x}|^2}.$$

Show that in this case, the equation of motion involves only the total mass M, and this implies that measurement of the orbital period and the semi-major axis of the system can lead to a determination of M only. How might the individual stellar masses m_1 and m_2 be found?

Exercise 18   (i) An isolated system consists of n particles $(n > 1)$ having position vectors $\mathbf{x}_1, \mathbf{x}_2, \ldots, \mathbf{x}_n$. Each pair of particles, i and j, possesses a mutual potential energy $U(|\mathbf{x}_{ij}|)$, where $\mathbf{x}_{ij} = \mathbf{x}_i - \mathbf{x}_j$, due to the interaction of the particles with each other. If the force exerted on particle i by particle j is given by

$$\mathbf{F}_{ij} = -\frac{dU(|\mathbf{x}_{ij}|)}{d|\mathbf{x}_{ij}|} \frac{\mathbf{x}_{ij}}{|\mathbf{x}_{ij}|},$$

write down $\mathbf{F}_{ji}$, the force exerted on particle j by particle i and confirm that Newton's third law (N3) is obeyed in the system.

(ii) Use the formula for the force in part (i) to work out the force $\mathbf{F}_{ij}$ in the following two cases:

(a) The electrostatic force between charged particles, for which

$$U(|\mathbf{x}_{ij}|) = \frac{Q_iQ_j}{4\pi\varepsilon_0|\mathbf{x}_{ij}|},$$

where Q_i and Q_j are the two charges in coulombs and ε_0 is a constant.

(b) The tension force between particles connected by a spring, for which

$$U(|\mathbf{x}_{ij}|) = \tfrac{1}{2}k|\mathbf{x}_{ij}|^2,$$

where k is the force constant of the spring.

Exercise 19   (i) An isolated system consists of *three* particles having masses m_1, m_2 and m_3, whose position vectors are $\mathbf{x}_1$, $\mathbf{x}_2$ and $\mathbf{x}_3$, and whose velocities are $\mathbf{v}_1$, $\mathbf{v}_2$ and $\mathbf{v}_3$. Write down expressions for the total momentum, $\mathbf{P}$, and total angular momentum, $\mathbf{J}$, of the system.

* (ii) If the forces between the particles obey N3 (cf. Section 5.5 Unit 3), show that $\mathbf{P}$ and $\mathbf{J}$ are both conserved quantities. [*Hint*: This problem is solved for a *two*-particle system in Section 5.6 Unit 3.]

Exercise 20   (i) An isolated system consists of three particles having masses m_1, m_2 and m_3. As in Exercise 18, the force exerted on particle i by particle j is given by

$$\mathbf{F}_{ij} = -\frac{dU(|\mathbf{x}_{ij}|)}{d|\mathbf{x}_{ij}|} \frac{\mathbf{x}_{ij}}{|\mathbf{x}_{ij}|},$$

where $U(|\mathbf{x}_{ij}|)$ is the potential energy due to the interaction between particles i and j. What is the total potential energy U, of the system?

(ii) Assuming that the kinetic energy of particle i is given by

$$T_i(|\mathbf{v}_i|^2) = \tfrac{1}{2}m_i|\mathbf{v}_i|^2,$$

write down the total kinetic energy T.

* (iii) Calculate $\dfrac{dU}{dt}$ and $\dfrac{dT}{dt}$ separately and show that, provided N2 is obeyed by the system, the total energy $E = T + U$ is conserved.

[*Hint*: The energy function for two particles is discussed in Section 5.3 Unit 3. Using for example Equation 15 as a guide, add further kinetic and potential terms to take account of the third particle. In calculating the time derivatives of U and T, apply the rule for composite functions (chain rule) of Section 5.4 Unit 3.]

4 Comprehension exercises for Block 2

Complete answers to these exercises may be found in the Glossary for Block 2 by looking up the word or phrase displayed in capital letters.

1 How is the CHARGE of a particle defined operationally, in both magnitude and sign?

2 State the law of ADDITIVITY OF CHARGE.

3 Write down the expression given by COULOMB'S LAW for the electrostatic force exerted on charged particle 1 by charged particle 2, when both are at rest in the same inertial reference frame.

4 If electrons are moving in a particular direction in a wire, what is the direction of the CURRENT?

5 EINSTEIN'S SPECIAL THEORY OF RELATIVITY is founded on two postulates, SR1 and SR2. State both postulates.

6 Explain how the ELECTRIC FIELD VECTOR, $\mathbf{E}(\mathbf{x}, t)$, at a position $\mathbf{x}$ and instant t in a given inertial reference frame is defined operationally.

7 'The ELECTRIC FIELD VECTOR differs in two inertial reference frames which are related by a uniform linear boost.' Is this statement true or false?

8 Define the ELECTRIC FLUX over a spherical surface of radius R. Is its value affected by charges (a) inside the sphere, (b) outside the sphere, or (c) both?

9 What is meant by an ELECTROMAGNETIC WAVE?

10 What is an EVENT in Einstein's special theory of relativity?

11 What is FARADAY'S EFFECT?

12 Define an INSTANTANEOUS REST FRAME for an accelerating particle in Einstein's special theory of relativity.

13 Explain what is meant by LENGTH CONTRACTION in Einstein's special theory of relativity.

14 Write down the expression for the force, $\mathbf{F}$, acting on a particle with charge q at an event $(\mathbf{x}, t)$ in an electromagnetic field, according to the LORENTZ FORCE LAW.

15 Does the LORENTZ FORCE LAW satisfy the principle of relativity? Is it form invariant under (a) the Galilean transformation, (b) the Lorentz transformation, or (c) both?

16 Give three examples of a LORENTZ INVARIANT, and explain what the term means.

17 Define the LORENTZ INVARIANT SPACETIME INTERVAL, ΔS, between two events in four-dimensional spacetime, observed in an inertial reference frame.

18 Write down the LORENTZ TRANSFORMATION between the spacetime coordinates of an event observed in an inertial frame of reference, and the spacetime coordinates of the same event observed in a second inertial frame.

19 Explain how the MAGNETIC FIELD VECTOR, $\mathbf{B}(\mathbf{x}, t)$, at a position $\mathbf{x}$ and instant t in a given inertial reference frame is defined operationally.

20 'The MAGNETIC FIELD VECTOR differs in two inertial reference frames which are related by a uniform linear boost.' Is this statement true or false?

21 MAGNETIC FIELD VECTORs are produced by (a) moving charges or currents, (b) time-varying electric fields, or (c) both. Which of the options (a), (b) or (c), if any, is correct?

22 'In Einstein's special theory of relativity the sum of the MASSes of two colliding particles is the same after the collision as before.' Is this statement true or false?

23 What is MAXWELL'S EFFECT?

24 What was the purpose of the MICHELSON–MORLEY EXPERIMENT? What was the result?

25 Outline the main steps necessary for setting up an OBSERVING SYSTEM. In what way does Einstein's special theory of relativity make the procedure simpler?

26 Does Einstein's special theory of relativity give rise to PARADOXES?

27 Define the PROPER TIME INCREMENT between two events on the world-line of a particle in Einstein's special theory of relativity.

28 Give a simple formula for the PROPER TIME INCREMENT, $\Delta\tau$, between two events on the world-line of a free particle.

29 How does the PROPER TIME INCREMENT between two events on the world-line of an accelerated particle compare to the proper time increment along the world-line of a free particle connecting the same two events?

30 Explain what is meant by the RELATIVISTIC DOPPLER SHIFT. In what way does it differ from the classical Doppler shift?

31 Write down an expression for the RELATIVISTIC ENERGY of a free particle in Einstein's special theory of relativity. In what way does it differ from the kinetic energy of a free particle in Newtonian physics?

32 Write down an expression for the RELATIVISTIC MOMENTUM of a free particle in Einstein's special theory of relativity.

33 A particle is moving with velocity u^1 along the **1**-axis of an inertial frame S. Use the RELATIVISTIC VELOCITY ADDITION formula to obtain an expression for the velocity, u'^1, of the particle along the **1**-axis of an inertial frame, S', which is moving with velocity v along the **1**-axis of S.

34 The RELATIVITY OF SIMULTANEITY in Einstein's special theory of relativity implies that an inertial frame can always be found in which two events are observed to occur simultaneously provided they are separated by (a) a time-like interval, (b) a space-like interval, or (c) either. Which, if any, of the options is true?

35 Define the REST FRAME for a free particle. Is it an inertial frame?

36 Explain what is meant by the statement that two events are separated by a SPACE-LIKE INTERVAL. Is it possible to find an inertial frame in which both events occur at the same position?

37 Write down an expression for the SPEED OF LIGHT in terms of the universal constants ε_0 and μ_0.

38 Explain the meaning of the term THRESHOLD ENERGY for a reaction.

39 Explain what is meant by TIME DILATION in Einstein's special theory of relativity.

40 Explain what is meant by the statement that two events are separated by a TIME-LIKE INTERVAL. Is it possible to find an inertial frame in which the events are simultaneous?

5 Consolidation exercises for Block 2

Introductory notes

The exercises in Section 5.1 are concerned with basic electromagnetism. There are two aspects to these: the creation of static electric and magnetic fields by distributions of charges and steady currents, and the forces exerted by these fields on charged particles. Apart from reinforcing the material of Unit 4, these exercises provide further practice in the manipulation and use of vectors. Where needed, the constant in Coulomb's law should be taken as: $(4\pi\varepsilon_0)^{-1} = 9 \times 10^9\,\mathrm{N\,m^2\,C^{-2}}$.

The short Section 5.2 explores some of the more immediate consequences of the two postulates of special relativity in an informal way. The exercises are based on the material of Unit 5, and are intended to be tackled without direct recourse to the Lorentz transformation. However, they anticipate the very important time dilation and length contraction effects which do not emerge in the course material until after the derivation of the Lorentz transformation in Unit 6.

The first exercise in Section 5.3 is based on the material of Section 2.1 Unit 6. It introduces simple spacetime diagrams and the idea of events as the intersections of world-lines. Again, the Lorentz transformation is not required. Most of the remaining exercises in this section and the next one are applications of the Lorentz transformation and associated Lorentz invariants to problems in relativistic kinematics and dynamics. A few of the exercises in Section 5.4 refer to processes involving elementary particles. You are not expected to know anything about these particles, beyond the fact that they possess linear momentum and energy. The examples have been chosen to illustrate the application of the conservation laws for these two quantities in physical situations where velocities comparable to that of light are commonplace.

Many of the exercises in Sections 5.2, 5.3 and 5.4 refer to two inertial observers, O and O′, whose reference frames are in *standard configuration*. This phrase denotes a set of conventions, detailed below, which is followed in order to avoid excessive repetition. Each observer may be considered to be located at the origin of his or her reference frame, and to control a complete observing system capable of assigning spacetime coordinates, $(ct, \mathbf{x})$, to remote, as well as nearby, events. The two reference frames are arranged so that:

(i) corresponding axes are parallel;

(ii) their clocks are synchronized, and their origins coincide at $t = t' = 0$; (This event is sometimes referred to as $\mathscr{E}_0$.)

(iii) O′ moves with constant speed v along the positive **1**-axis of O. (In some exercises, a specific numerical value is given for v.)

Where only one space dimension is involved, it is referred to as x, rather than $x^{\mathbf{1}}$, and in this case the Lorentz transformation takes the form

$$x' = \gamma\left(x - \frac{v}{c}ct\right) \qquad ct' = \gamma\left(ct - \frac{v}{c}x\right),$$

where $\gamma = 1/\sqrt{1 - v^2/c^2}$ (cf. Equations 18, 19 and 20 of Unit 6). These equations give the spacetime coordinates assigned to an event by O′ in terms of the spacetime coordinates assigned to the same event by O.

Interchanging the two observers and replacing v by $-v$, we obtain the inverted equations:

$$x = \gamma \left(x' + \frac{v}{c} ct' \right) \qquad ct = \gamma \left(ct' + \frac{v}{c} x' \right)$$

(cf. Equations 21 and 22 of Unit 6).

If the problem is such that the discussion cannot be restricted to one spatial dimension, we continue to use x^1, x^2 and x^3. In this case, the appropriate form of the Lorentz transformation is that given in Equations 23, Unit 6.

5.1   Unit 4

Exercise 21   A charge q_1 located at $\mathbf{x}_1 = (0.4, -0.6, 1)$ m is acted on by a force $\mathbf{F}_{12} = (-9, 12, -36)$ N due to a charge q_2 located at $\mathbf{x}_2 = (-0.2, 0.2, -1.4)$ m. If $q_1 = 2.5 \times 10^{-4}$ C, what is the value of q_2?

Exercise 22   Five charges are located as follows:
(a) 3×10^{-4} C at $(-0.6, -0.6, 0.6)$ m
(b) 6×10^{-4} C at $(0.5, -0.5, 0.5)$ m
(c) 2.5×10^{-4} C at $(0.7, 0, -0.7)$ m
(d) -5×10^{-4} C at $(0, 0, 0)$ m
(e) -2×10^{-4} C at $(-0.8, 0.6, 0.1)$ m.

Calculate the *average* value of the radial component of the total electric field due to these charges on the surface of an imaginary sphere of radius 1 m centred at $(0, 0, 0)$.

Exercise 23   (i) Consider the electric field, $\mathbf{E}$, of a very long, thin plastic rod which carries a uniform density of charge, λ, along its length. For simplicity, assume the rod occupies the entire **1**-axis of a reference frame, from $x^1 = -\infty$ to $x^1 = \infty$. Would it be possible for the electric field lines of the rod to be in the form of circular loops, like the magnetic field in Figure 20 Unit 4?

(ii) The field must be invariant under static translations of the coordinates along the **1**-axis, so that the point, P, at which the field is to be determined can be placed in the **2,3**-plane without loss of generality.

Consider two very small portions of the rod, each of length Δx, situated at $x^1 = \pm x$, where $x \geqslant 0$. Using Equation 4 Unit 4, which gives the electric field due to two point charges, show that their combined contribution to the electric field at P is

$$\frac{\lambda \, \Delta x \, \mathbf{r}}{2\pi\varepsilon_0 (r^2 + x^2)^{3/2}},$$

where $\mathbf{r} = (0, x^2, x^3)$ and $r = |\mathbf{r}| = \sqrt{(x^2)^2 + (x^3)^2}$ is the perpendicular distance from P to the rod. Notice that there is no electric field parallel to the rod. The field due to the whole rod can be calculated by summing the contributions of all such pairs of elements, i.e. by integrating from $x = 0$ to $x = \infty$. The result turns out to be $\mathbf{E} = E\mathbf{r}/r$, where $E = \lambda/2\pi\varepsilon_0 r$.

*   (iii) It is stated in Section 2.3 Unit 4 that if an imaginary sphere is drawn in a region containing any distribution of charges, then $\Phi = Q/\varepsilon_0$, where Q is the total charge within the sphere. In this equation, Φ is the total electric flux over the sphere, and is given as the average *radial* component of the electric field multiplied by the surface area of the sphere. This would be a strange physical law if it were true only for imaginary spheres! Assume that the statement is true for a closed surface of *any* shape, but that Φ is to be interpreted more generally as the average value of the *normal* (i.e. perpendicular to the surface) component of the electric field multiplied by the area of the surface. Consider a cylindrical surface of arbitrary length, drawn symmetrically about the charged rod, and assume that the field is of magnitude E and in the direction of the unit vector $\mathbf{r}/r$. Show that E must be given by the expression quoted in part (ii).

Exercise 24   A small positive test charge located at the point $(1, 0, 0)$ and moving with velocity $(1, -1, 2)$, is subject to static electric and magnetic fields. Write down a vector giving the *direction* of the acceleration of the test charge when the source of the field is each of the following:

(i) A stationary negative charge at $(0, 1, 0)$.

(ii) A small bar magnet fixed with its centre at the origin and its N–S axis lying along the **3**-axis. The north pole of the magnet is located on the positive **3**-axis. (The magnetic field lines of a bar magnet are illustrated in Figure 23 Unit 4.)

(iii) A very long straight rod of insulating material carrying a uniform density of positive charge along its length, lying stationary along the **2**-axis.

(iv) A very long straight neutral wire lying stationary along the **3**-axis, in which there is a steady flow of electrons in the negative **3**-direction. (The magnetic field lines are shown in Figure 20 Unit 4.)

(v) A small neutral loop of wire lying stationary in the **2,3**-plane with its centre at the origin, in which there is a steady circulation of electrons in a clockwise sense when viewed from the point $(1, 0, 0)$. (The magnetic field lines for this case resemble those of the short solenoid in Figure 24 Unit 4.)

Exercise 25   (i) A particle has mass m and carries charge q. Initially it is located at $\mathbf{x} = \mathbf{0}$ and moving with velocity $\mathbf{v} = (0, 0, v)$, with $v \ll c$, when it enters a narrow region of static electric and magnetic fields between $x^3 = 0$ and $x^3 = d$. The fields are given by $\mathbf{E} = (E, 0, 0)$ and $\mathbf{B} = (B, 0, 0)$, where E and B are both constant. Write down the acceleration vector, $\mathbf{a}$, of the particle in terms of its velocity components v^2 and v^3. Why does $\mathbf{a}$ not depend on v^1?

(ii) Evaluate $\mathbf{a}$ at $\mathbf{x} = \mathbf{0}$, and, assuming that it does not change appreciably between $x^3 = 0$ and $x^3 = d$, show that

$$v^2 \approx qBd/m, \quad x^1 \approx \frac{qEd^2}{2mv^2} \quad \text{and} \quad x^2 \approx \frac{qBd^2}{2mv},$$

when $x^3 = d$. [*Hint*: Use the formulae for uniformly accelerated motion in Section 4.5 Unit 1.]

* (iii) Show also that the approximation used in part (ii) will give accurate results if $d \ll mv/qB$.

* (iv) A large number of particles of the same kind, with a wide range of velocities, are collimated into a narrow beam and passed through the apparatus. Their arrival at $x^3 = d$ is recorded on a photographic plate. Show that they produce a parabolic track on the plate, and that other kinds of particles, with different values of q/m, will lie on different parabolas.

5.2   Unit 5

Exercise 26   A long neutral wire lies stationary along the **1**-axis of the reference frame of an observer O, and carries a steady current I in the negative direction. A test charge of small positive magnitude q is *at rest* at the point $\mathbf{x}' = (0, 0, r)$ in the reference frame of a second observer O'. The reference frames of O and O' are in standard configuration.

(i) Referring to Section 3.3 Unit 4 for details, write down the magnetic field, **B**, at a point $\mathbf{x} = (x, 0, r)$ due to the current. Hence show that the force, **f**, exerted by the magnetic field on the test charge is given by $\mathbf{f} = (0, 0, f)$, where

$$f = \frac{\mu_0 q I v}{2\pi r}.$$

(ii) Suppose we make the assumption, embodied in Equations 2a and 2b of Unit 5, that the Lorentz force law is valid for *any* inertial observer, although the electric and magnetic fields have different values in different frames. The wire is observed by O to be electrically neutral, so that the electric field, **E**, in his frame is zero. It follows that the force of repulsion between the wire and the test charge, which is observed by both O and O', is attributable in the O frame to the magnetic field only. Explain how the situation differs in the O' frame. In particular, why would we expect the wire to be electrically *charged* in the O' frame?

(iii) In Exercise 23, the electric field due to an infinite line charge of constant density λ, lying along the **1**-axis, is given as

$$\mathbf{E} = \frac{\lambda \mathbf{r}}{2\pi \varepsilon_0 r^2},$$

where $\mathbf{r} = (0, x^2, x^3)$ and $r = |\mathbf{r}|$. Use this result to show that if the wire carries a line charge of density λ in the O' frame, the force $\mathbf{f}'$, acting on the test charge in this frame is given by

$$\mathbf{f}' = \frac{\lambda c^2}{I v} \mathbf{f}.$$

[*Hint*: You will need the relation $\mu_0 \varepsilon_0 c^2 = 1$ given in Section 4.3 Unit 4.]

Consideration of this problem is resumed in Exercise 32.

Exercise 27   (i) Observers O and O', whose reference frames are in standard configuration, observe the motion of a light signal. The signal is emitted at $t = t' = 0$ from the origin of the O' frame towards the point $\mathbf{x}' = (0, d, 0)$, where it is reflected back towards $\mathbf{x}' = \mathbf{0}$. Its time of arrival at $\mathbf{x}' = \mathbf{0}$ is recorded by O' as $t' = T'$ and by O as $t = T$. Draw sketches of the motion of the light signal in both reference frames and show directly that the total distance travelled by the light signal according to O is

$$\sqrt{4d^2 + v^2 T^2}.$$

What total distance is travelled by the light signal according to O'?

(ii) Use the results of part (i) to obtain an expression for the velocity of light according to each observer, and equate both expressions to c, in accordance with SR2 introduced in Section 3.3 Unit 5. Eliminate the distance d between the two equations, and hence obtain the time dilation formula (Section 3.2 Unit 6) $T = \gamma T'$, where $\gamma = 1/\sqrt{1 - v^2/c^2}$.

(iii) How are T and T' related in the Newtonian world-view?

Exercise 28   (i) The reference frames of two observers, O and O′, are in standard configuration. A rod, AB, lies at rest in the O frame from $x = x_A = 0$ to $x = x_B = L$. According to O, O′ traverses the rod from A to B in a time interval T. If O′ records the time he takes to traverse the rod as T', and calculates its length by multiplying this time by the relative velocity, show that his answer is

$$L' = \left(\frac{T'}{T}\right) L.$$

(ii) Use this result, together with the time dilation formula to deduce the length contraction formula (Section 3.1 Unit 6.) $L' = L/\gamma$, where $\gamma = 1/\sqrt{1 - v^2/c^2}$.

Exercise 29   An object is moving with uniform speed away from an inertial observer, O, along his positive **1**-axis. At time $t = 5\,\text{s}$, O emits a short light-pulse, which, after reflection from a mirror on the object, returns to him at $t = 8\,\text{s}$. While waiting for the first pulse to return, he emits a second pulse at $t = 7\,\text{s}$, which is also reflected from the object, and returns at $t = 12\,\text{s}$. What is the equation of the world-line of the object in the O frame?

5.3   Unit 6

A brief tutorial: 'Which Lorentz transformation to use?'

A large number of the problems you are likely to get in special relativity require, for their solution, the Lorentz transformation from (t, x) to (t', x') corresponding to two inertial frames in standard configuration. There are, of course, two such transformations, one the reverse of the other. One transformation expresses (t', x') in terms of (t, x) and the inverse expresses (t, x) in terms of (t', x'). But which transformation should one choose? Before considering this question with two examples, we start by doing what you should do when confronted by such a problem: write down the Lorentz transformation equations. Here they are with some blanks between square brackets [   ] for you to fill in (referring to Unit 6 if necessary):

$$x = [\quad]\,([\quad]x'\,[\pm?\quad][\quad]t')$$

$$t = [\quad]\,([\quad]t'\,[\pm?\quad][\quad]x')$$

$$x' = [\quad]\,([\quad]x\,[\pm?\quad][\quad]t)$$

$$t' = [\quad]\,([\quad]t\,[\pm?\quad][\quad]x)$$

There are similar equations in terms of $\Delta x, \Delta t$ etc. The question to ask yourself when considering which of these to use in a particular problem is: what stays constant? Think back to two important cases: (i) *Time dilation*: Consider a uniformly moving clock; does it go slow compared to a stationary clock? Consider the clock in the frame in which it is at rest, the primed (moving) frame. In this frame, two ticks will be two events separated by time interval $\Delta t'$ but zero interval in position, $\Delta x'$. So, we use the equation in which putting $\Delta x' = 0$ simplifies things, i.e. the equation for Δt in terms of $\Delta t'$ and $\Delta x'$, the inverse Lorentz transformation. This gives us the well-known time dilation expression $\Delta t = \gamma \Delta t'$, or even $\Delta t = \gamma \Delta \tau$ since we have a name for and symbol for the time on a clock carried along stationary with the observer, proper time τ. And, since $\gamma \geqslant 1$, you can see that the elapsed time for a stationary observer exceeds the elapsed proper time. Of course, you *can* derive the results from the 'direct' Lorentz transformation, but you have to calculate Δx.

(ii) *Length contraction*: If you have something which has length $\Delta x' = L$ in its (moving) rest frame, what length is it in your (un-primed) rest frame? Well, by length in your frame, you mean length at one instant of time ($\Delta t = 0$), so now the equation to use is the one which is simplified by the fact that $\Delta t = 0$. This is the equation for $\Delta x'$ in terms of Δx and Δt, i.e. the direct Lorentz transformation. It gives $\Delta x' = \gamma \Delta x$, i.e. $\Delta x = L/\gamma$, the length contraction expression.

So, we have length *contraction* and time *dilation,* and the following notion: the transformation to use is the one which is simplified by the fact that some time interval is zero (because something is simultaneous in one frame or the other) or some spatial interval is zero (two events at the same place in one frame or the other.)

Exercise 30   (i) A light signal is observed by two observers, O and O′, whose reference frames are in standard configuration. The signal is emitted from the origin of the O frame at $t = t_1 > 0$ (event $\mathscr{E}_1$) and received at the origin of the O′ frame at $t' = t'_2$ (event $\mathscr{E}_2$). Suppose event $\mathscr{E}_1$ occurs at $t' = t'_1$ by the O′ clock and event $\mathscr{E}_2$ occurs at $t = t_2$ by the O clock. Both observers agree that the light signal travels with velocity c, and each makes a sketch, similar to Figures 3 and 19 (Unit 6), showing the world-line of the *other* observer and that of the light signal. Reproduce both their sketches.

(ii) Write down the equations for the world-lines of O′ and the light signal in the O sketch, and hence show that $t_2 = t_1/(1 - v/c)$.

In a similar way, use the equations of the world-lines in the O′ sketch to show that $t'_1 = t'_2/(1 + v/c)$.

(iii) According to the time dilation formula, what would you expect the ratio t_2/t'_2 to be? Use this ratio together with the results of part (ii) to obtain the ratio t_1/t'_1. Is your answer for this ratio consistent with the time dilation formula? [*Hint*: You may find it helpful to think of $\mathscr{E}_0, \mathscr{E}_1$ and $\mathscr{E}_0, \mathscr{E}_2$ as separate pairs of events, where $\mathscr{E}_0$ is $x = x' = 0, t = t' = 0$.]

Exercise 31   (i) A train enters a tunnel of length 56 m at a uniform speed of $\frac{45}{53}c$. Let the disappearance of the rear of the train at the mouth of the tunnel be event $\mathscr{E}_0$ and the reappearance of the engine at the far end of the tunnel be event $\mathscr{E}_1$. An observer O, stationed at the tunnel mouth asserts that these two events are simultaneous. A second observer O′, is located at the rear of the train, and the reference frames of O and O′ are in standard configuration. If $\mathscr{E}_0$ occurs at $t = t' = 0$, at what time on the O′ clock does $\mathscr{E}_1$ occur? How long is the train?

(ii) Draw a spacetime diagram from the viewpoint of O, taking the space axis in the direction of motion of the train, which shows the world-lines of both ends of the train and both ends of the tunnel. Indicate the events $\mathscr{E}_0$ and $\mathscr{E}_1$ on your diagram.

Using the results from part (i) concerning the length of the train and the time of $\mathscr{E}_1$, draw another spacetime diagram from the viewpoint of O', showing the same four world-lines and two events. Indicate on your diagram the length of the tunnel as observed by O'. What is this length?

* (iii) Where is O' at the moment at which he *sees* the engine appear at the far end of the tunnel? Let this event be $\mathscr{E}_2$. Calculate the time at which $\mathscr{E}_2$ occurs according to O, and according to O', and mark it on the diagrams you have drawn in part (ii).

* (iv) Later, the train enters another tunnel and O' sees the engine appear at the far end of the tunnel just as he reaches its mouth. What is the length of this second tunnel?

Exercise 32 This question is intended as a sequel to Exercise 26.

(i) The current in the wire is produced by electrons, each carrying charge $-e$ which move with velocity u along the positive **1**-axis. The bulk of the wire consists of positive ions, each having charge e. There are n positive ions per metre of wire. Thus in the O frame the wire consists of two infinite line charges, one with positive density $\lambda^+ = ne$ which is stationary, and another with negative density λ^- (measured in its own rest frame) moving with velocity u along the positive **1**-axis. The wire is electrically neutral overall, but the moving line charge is subject to the length contraction effect, as discussed in Section 3.4 Unit 6. Show that this implies

$$\lambda^- = -ne\sqrt{1 - u^2/c^2}.$$

(ii) Show that there is an imbalance between the positive and negative charge densities when the wire is observed by O' (referring to Exercise 26), and that the net charge density is

$$\lambda = ne\left[\frac{1}{\sqrt{1 - v^2/c^2}} - \frac{\sqrt{1 - u^2/c^2}}{\sqrt{1 - u'^2/c^2}}\right],$$

where u' is the velocity of the moving electrons in the wire relative to O'.

* (iii) Evaluate λ approximately for $u \ll c$ and $v \ll c$, by writing $u' \approx u - v$ and using the expansions

$$\sqrt{1 - x^2} \approx 1 - \tfrac{1}{2}x^2 \qquad 1/\sqrt{1 - x^2} \approx 1 + \tfrac{1}{2}x^2$$

for the square root factors. Hence show that $\lambda \approx neuv/c^2$.

* (iv) An infinite line charge of density ne moving parallel to its length with velocity u, constitutes a current of magnitude $I = neu$ (cf. Equation 9 Unit 4). Thus the linear charge density on the wire in the O' frame can be written as $\lambda \approx Iv/c^2$. Use this result, together with the last part of Exercise 26, to show that $\mathbf{f}' \approx \mathbf{f}$, i.e. that the force on the test charge observed by O' is approximately equal to that observed by O.

Exercise 33 (i) The electric field associated with a plane light wave is represented in the reference frame of an observer O by $\mathbf{E} = (0, 0, E)$, where

$$E = E_0 \cos[2\pi(x/\lambda - ft)].$$

Here λ and f are the wavelength and frequency of the light, respectively, and satisfy the relation $\lambda f = c$. In the reference frame of a second observer O', which is in standard configuration with that of O, the electric field associated with the same plane light wave is $\mathbf{E}' = (0, 0, E')$, where

$$E' = E_0' \cos[2\pi(x'/\lambda' - f't')].$$

By applying the Lorentz transformation to the expression $x/\lambda - ft$, rewrite it in terms of x' and t', and hence obtain λ' and f' in terms of λ, f and v/c.

(ii) Show that $\lambda' f' = c$, and explain why this result could have been expected.

Exercise 34 The reference frames of two observers O and O$'$ are in standard configuration with relative velocity $v = 5c/13$. At time $t_1 = 3/c$, O sends a light signal in the direction of O$'$. Let the departure of this light signal be event $\mathscr{E}_1$. On receipt of the signal, which we call event $\mathscr{E}_2$, O$'$ immediately dispatches a rocket towards O, which travels at velocity $-4c/5$ relative to O$'$. The arrival of the rocket at O, at time t_3 by his clock, is event $\mathscr{E}_3$.

(i) Calculate the spacetime coordinates (ct_2, x_2) of event $\mathscr{E}_2$ as observed by O. At what time by the O$'$ clock does this event occur?

(ii) Calculate the velocity of the rocket relative to O, and use it to obtain t_3. [*Hint*: The transformation of velocities between inertial observers is discussed in Section 3.3 Unit 6.]

Draw a spacetime diagram showing the complete sequence of events from the viewpoint of O.

(iii) Assuming that the rocket travelled with O$'$ on the outward journey, what is the duration of the round-trip from O and back to O according to a clock aboard the rocket? How does your answer compare with t_3? Does your result agree with the discussion of the twin 'paradox' in Section 4.2 Unit 6?

5.4 Unit 7

* **Exercise 35** Consider observers O and O$'$ in standard configuration. Draw a spacetime diagram from the viewpoint of O, showing the ct'-axis $(x' = 0)$ and the x'-axis $(ct' = 0)$ of observer O$'$. [*Hint*: Lines of constant position and time for O$'$ are discussed in Section 4.1 of Unit 7, and illustrated in Figure 9 of that unit.]

A straight rod, AB, lies stationary along the x-axis of the O frame from $x = x_A = 0$ to $x = x_B = L$.

Indicate the following on your diagram:

(i) the world-lines of A and B;

(ii) a segment of the x'-axis representing the length L', of the rod as determined by O$'$;

(iii) a segment of the ct'-axis representing cT', where T' is the time taken by O$'$, as recorded on his clock, to travel the length of the rod from A to B.

Hence show that $L' = vT'$, where v is the velocity of O$'$ relative to O.

Exercise 36 (i) Write down the Lorentz transformation giving the spacetime coordinates, (ct', x'), of an event as determined by O$'$ in terms of its coordinates, (ct, x), determined by O. Assume the two reference frames are in standard configuration with relative velocity $v = 3c/5$.

(ii) The motions of two particles, P and Q, are observed by O and O$'$. Particle P is released from $x = 0$ at $t = 0$, and travels in the positive x-direction with speed $4c/5$ relative to O. Particle Q is released from $x' = 4$ at $t' = 0$ and travels in the negative x-direction with speed $c/2$

relative to O. What are the equations of the world-lines of P and Q according to O, and when and where do they collide?

Where and when do P and Q collide according to O′?

Exercise 37   (i) Two events, $\mathscr{E}_a$ and $\mathscr{E}_b$, are observed by O to occur at the *same place*, and separated by a time interval of 15 s. For a second observer O′, whose reference frame is in standard configuration with that of O, the time separation between $\mathscr{E}_a$ and $\mathscr{E}_b$ is 17 s. Calculate the spatial separation of the two events according to O′, and write down their spacetime separation, $(c(t_b' - t_a'), x_b' - x_a')$. [*Hint*: Apply Theorem I, Section 3.2 Unit 7.]

(ii) In a repeat experiment, involving the same two events and with the same relative motion, O and O′ arrange $\mathscr{E}_a$ to occur at the intersection of their world-lines. Use the result of part (i) to write down the spacetime coordinates of $\mathscr{E}_b$ according to O′. In this experiment, O′ allows 7 s to pass after $\mathscr{E}_a$ and then instantaneously reverses his direction of motion, so that he travels back towards O at constant velocity. If his new world-line intersects that of O at $\mathscr{E}_b$, calculate the total proper time recorded on the O′ clock between $\mathscr{E}_a$ and $\mathscr{E}_b$. Is your result in agreement with Theorem II, Section 5.3 Unit 7?

(iii) What is the maximum time O′ could have waited before reversing his motion, and still kept his appointment with O at $\mathscr{E}_b$? Explain your reasoning.

Exercise 38   Observers O and O′ who are both familiar with Newtonian mechanics, are trying to establish formulae for the momentum, **p**, and energy, E, of a particle, which are valid for arbitrarily high velocity. They believe that **p** is proportional to the velocity, **v**, of the particle, and that $(p^0, \mathbf{p})$ forms a four-vector, where $p^0 = E/c$. By requiring this four-vector to transform under any Lorentz transformation in exactly the same manner as $(x^0, \mathbf{x})$ with $x^0 = ct$, O and O′ are led to the assignments

$$\mathbf{p} = \frac{m\mathbf{v}}{\sqrt{1 - |\mathbf{v}|^2/c^2}} \quad \text{and} \quad p^0 = \frac{mc}{\sqrt{1 - |\mathbf{v}|^2/c^2}}.$$

Reconstruct their argument by following the steps listed below.

(i) Restricting the argument to one spatial dimension, and assuming that O and O′ have arranged their reference frames in standard configuration, write down the Lorentz transformation giving (x^0, x) in terms of (x'^0, x'). Replace the spacetime variables, in the transformation you have written down, by momentum variables in order to obtain equations relating (p^0, p) to (p'^0, p').

(ii) Apply this transformation to a particle which is *at rest* in the O′ frame, and hence show that

$$p = \gamma p'^0 v/c \quad \text{and} \quad p^0 = \gamma p'^0.$$

(iii) By comparing p with the Newtonian form in the non-relativistic limit $(v \approx 0)$, show that $p'^0 = mc$, where m is the mass of the particle. Substitute for p'^0 in the results of part (ii) and show that

$$p = \gamma mv \quad \text{and} \quad p^0 = \gamma mc.$$

(iv) Show that the results obtained in part (iii) are consistent with the assignments made by O and O′.

Exercise 39   (i) A negative π meson or 'pion', (π^-), of mass m_π and momentum $\mathbf{p}_\pi$, collides with a *stationary* proton (p^+) of mass m_p. Show that the time and space components of the four-momentum for the entire system are given by

$$P^0 = E_\pi/c + m_\mathrm{p}c \quad \text{and} \quad \mathbf{P} = \mathbf{p}_\pi,$$

where E_π is the energy of the pion. Write down an expression for E_π in terms of m_π and $\mathbf{p}_\pi$.

(ii) Explain what is meant by the statement that the expression

$$(P^0)^2 - |\mathbf{P}|^2$$

is a Lorentz invariant, and calculate its value in terms of E_π, m_π and m_p.

(iii) Calculate the total energy of the system in the centre-of-momentum reference frame, which is defined as that in which the total momentum of the system is zero.

(iv) Use the result of part (iii) to calculate the minimum value of E_π, i.e. the *threshold energy* for the reaction $\pi^- + p^+ \longrightarrow \Lambda^0 + \mathrm{K}^0$, in which two 'strange particles' are produced by the collision. What is the speed of the pion relative to the target proton at this threshold energy? [The energy equivalents, mc^2, for the masses of the π^-, p^+, Λ^0 and K^0 particles are 140, 938, 1116 and 498 MeV respectively, where 1 MeV (megaelectronvolt) is the unit of energy commonly used in nuclear physics. It is equivalent to the kinetic energy acquired by an electron which accelerates through a distance of 1 m in an electric field of strength $10^6 \, \mathrm{N\,C^{-1}}$. The value in SI units is given on the back cover, but is not required. It is perfectly acceptable to give answers in MeV if the data are in MeV.]

Exercise 40   (i) Use the formulae obtained in Exercise 38, to show that the velocity and energy of a particle of mass m and momentum $\mathbf{p}$ are given by

$$\mathbf{v} = \frac{\mathbf{p}c}{\sqrt{m^2c^2 + |\mathbf{p}|^2}} \quad \text{and} \quad E = c\sqrt{m^2c^2 + |\mathbf{p}|^2}.$$

(ii) What is the velocity of a *massless* particle with momentum $\mathbf{p}$? What is its energy?

(iii) Show that the velocity of a particle of any mass can be written as $\mathbf{v} = \mathbf{p}c^2/E$.

Exercise 41   (i) A massless particle with energy E moves in the **1,2**-plane of the O reference frame, on a path making an angle θ with the positive **1**-axis. Write down the time and space components of its momentum, p^0 and $\mathbf{p}$.

(ii) Recalling that $(p^0, \mathbf{p})$ transforms like $(ct, \mathbf{x})$ under a Lorentz transformation, show that the energy, E', of the particle as determined by O', whose reference frame is in standard configuration with that of O, is

$$E' = E\gamma \left(1 - \frac{v}{c}\cos\theta\right),$$

where $\gamma = 1/\sqrt{1 - v^2/c^2}$.

(iii) Show also that the path of the particle makes an angle θ' with the O' **1**-axis, where

$$\tan\theta' = \frac{\sin\theta}{\gamma(\cos\theta - v/c)}.$$

[*Hint*: More than one spatial dimension is involved here, so the appropriate form of the Lorentz transformation is given in Equations 23 of Unit 6.]

Exercise 42   (i) A neutral π meson (pion) decays into two photons (γ-rays), i.e. $\pi^0 \longrightarrow \gamma + \gamma$. The pion has mass m_π and the photons are massless. Consider this process in the rest frame of the pion, and write down the energy of each photon in terms of m_π. What can be said about the directions in which the photons emerge?

* (ii) Consider a possible process in which a positron-electron pair is produced from a single high-energy photon, i.e. $\gamma \longrightarrow e^+ + e^-$. The photon is massless, and the two charged particles each have mass m_e. Let the energies and momenta of the positron and electron be $E_1, \mathbf{p}_1$ and $E_2, \mathbf{p}_2$ respectively. Show that the reaction could occur only if

$$\frac{c|\mathbf{p}_1 + \mathbf{p}_2|}{E_1 + E_2} = 1.$$

[*Hint*: What is the ratio $c|\mathbf{p}|/E$ for a photon? Exercise 40 parts (ii) and (iii) may be helpful.]

* (iii) Show that this ratio is strictly less than 1 and, in consequence, 'pair production' by this means is impossible. Explain what prevents the process from taking place.

6   Spacetime diagrams tutorial

It is often useful to draw a simple spacetime diagram — ct against x — to display events for a given inertial frame of reference. But the principle of relativity says that all inertial frames are equivalent. So it is important to be able to say how the same events are related by two inertial observers. [This is the content of the Lorentz transformation, given *mathematically* by Equations 23 of Unit 6.] Since we practically always use the standard configuration in which the origins coincide at time zero, it is possible to draw a second spacetime diagram, corresponding to a different frame, overlaid on the first and lined up on the origin. This is explained in general terms in Section 4.1 of Unit 7. Many people have found that a very detailed use of this technique, on specific examples, has greatly improved their understanding of the way the Lorentz transformation works. Dr Bob Zimmer, who was the author of the Glossary for S357, has developed a set of diagrams which he has tested in a number of tutorials and talks, and which are now published.

R. S. Zimmer *Getting the picture in special relativity*, IET, The Open University ISBN 0-9519683-0-0

He has generously allowed us to use some of the diagrams and has written the following text to incorporate them into this Unit. We include this in the hope that some of you may find it a useful aid, used in conjunction with Units 6 and 7.

Axes in spacetime diagrams

Spacetime diagrams let us read off the coordinates of events in spacetime in our own inertial frame of reference O, in comparison to the coordinates of the same events in spacetime as measured by observers in another inertial frame of reference O' moving relative to our own. The diagrams do this by superimposing on a single graph the axes and scales of the two coordinate systems which represent the two frames of reference.

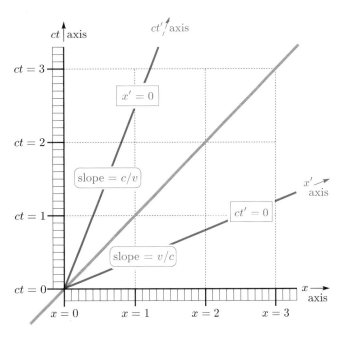

Figure 2   Axes in spacetime diagrams. For the case shown $v/c = 0.4$.

The axes which we use in drawing spacetime diagrams adopt the convention used in the units. We define x in the usual way as our space variable. But instead of defining t as our time variable, we multiply it by the velocity of light to produce a new time variable, ct. The reason: in terms of the variables x and ct, the Lorentz transformation equations take on a conveniently symmetric form. This symmetry shows up in the spacetime diagram in Figure 2.

In this diagram, the x-axis is our space axis, and the ct-axis is our time axis. It is the convention in spacetime diagrams that we draw the axes representing our own inertial frame of reference O as standard Cartesian axes — vertical and horizontal on whatever page we are using. It is always the axes representing the inertial frame O′ in which the other observer is at rest that we draw as tilted.

Tilted by how much? By definition, our x-axis (our space axis) is the line along which $ct = 0$, and our ct-axis (our time axis) is the line along which $x = 0$. In just the same way, for the other observer the $x′$-axis (her space axis) is the line along which $ct′ = 0$, and the $ct′$-axis (her time axis) is the line along which $x′ = 0$. To see what these other axes look like in terms of own own, we first set $ct′ = 0$ and then $x′ = 0$ in the Lorentz transformation equations. The result is as shown in the spacetime diagram of Figure 2: the other observer's space axis, the $x′$-axis, appears to us as a straight line of slope v/c through the origin; her time axis, the $ct′$-axis, appears to us as a straight line of slope c/v through the origin.

As v/c approaches the value zero, the axes representing the other observer's inertial frame become indistinguishable to us from those representing our own. As v/c approaches the value one, the axes representing her inertial frame for us squeeze closer and closer together, to become indistinguishable to us from the line of slope equal to one. Figure 2 has been drawn for a value of $v/c = 0.4$, to make both sets of axes most easily visible.

Scales in spacetime diagrams

Once we have drawn the axes of a spacetime diagram, all that remains is to get the scales right. The easiest way is to look at intercepts. If we ask where the line along which $x′ = 1$ intercepts our x-axis, we can find out by setting $x′ = 1$ and $t = 0$ in the Lorentz transformation equations. It intercepts our x-axis at $x = 1/\gamma$, where γ is the famous factor $\gamma = 1/\sqrt{1 - (v/c)^2}$, a quantity which is greater than 1.

We can do the same for the lines of $x′ = 2$, $x′ = 3$ and so on. The result is a series of intercepts on our x-axis which shows us how far apart to draw these lines.

Equally, if we ask where the line along which $ct′ = 1$ intercepts our ct-axis, we can find out by setting $ct′ = 1$ and $x = 0$ in the Lorentz transformation equations. It intercepts our ct-axis at $ct = 1/\gamma$.

We can do the same for the line of $ct′ = 2$, $ct′ = 3$ and so on. The result is a series of intercepts on our ct-axis which are exactly the same distance apart as the intercepts on our x-axis.

When we draw the lines of constant values of $x′$ together with the lines of constant values of $ct′$, as shown in Figure 3, we obtain a grid which displays the change in scale from our coordinate system to that of the other observer.

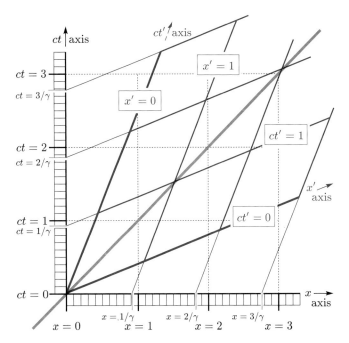

Figure 3   Scales in spacetime diagrams. Once again $v/c = 0.4$.

This diagram completely defines the (ct', x') coordinate system in terms of our own (ct, x) coordinate system. It displays for immediate visual grasp all of the information which is contained algebraically within the Lorentz transformation equations.

Reverse transformations

Of course, the other observer can do exactly the same as we have done, and the result as she would draw it appears in Figure 4. She will draw her own axes as the Cartesian ones, perpendicular on the page, and our axes as the tilted ones.

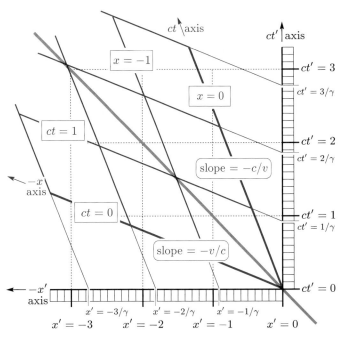

Figure 4   Reverse transformation for the case $v/c = 0.4$.

So long as she has agreed with us to use the same notation and to refer to the same pair of inertial frames in the same standard configuration, then there is only one difference between her and us: objects at rest in her inertial frame of reference move away from us with x-component of velocity $+v$ along our space axis, whereas objects at rest in our inertial frame move away from her with x-component of velocity $-v$ along her space axis. So the slopes of our axes as seen in her coordinate system are negative rather than positive, and the diagram which she then draws is the exact reverse of ours.

It is on account of the initial agreement between her and us — to use the same notation and to refer to the same pair of inertial frames in the same standard configuration — that we know what they see, we know that they know what we see, and we know that everyone knows that everyone knows.

The meaning of simultaneity

The very existence of the axes in a spacetime diagram makes a major statement about the meaning of simultaneity.

In ordinary life, we take it for granted that we can talk about the same measurable instant in time at all measurable positions in space. In the spacetime diagram of Figure 5, the line $ct = 1$ represents, within our own inertial frame of reference, one measurable instant in time at all measurable positions in space. The line $ct = 2$ represents another, and so on. Two events that lie together on any one of these lines are said to be simultaneous. The meaning of simultaneity — *within our own inertial frame of reference* — is the same as in ordinary life.

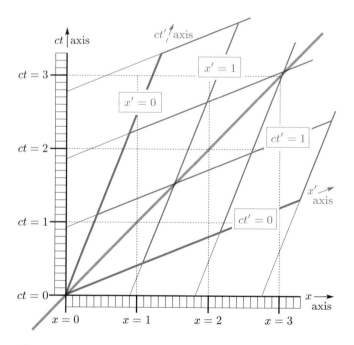

Figure 5   The meaning of simultaneity.

Equally, the line $ct' = 1$ represents, within the other observer's inertial frame, one measurable instant in time at all measurable positions in space. The line $ct' = 2$ represents another. Within the other observer's frame, two events that lie together on any one of these lines are said to be simultaneous. Again, the meaning of simultaneity — within their inertial frame — is the same as in ordinary life.

In short, simultaneity in special relativity is nothing mysterious. As in ordinary life, it means that two events at different measured positions in space happen at the same measured instant in time. The only difference from ordinary life is that *we have to specify the particular inertial frame of reference in which the measurements are being made.*

The relativity of simultaneity

We have seen that within any one inertial frame of reference, simultaneity has the same meaning in special relativity as it has in ordinary life: events take place at different places at the same time.

But what happens if we compare what is observed in one inertial frame, with what is observed in another? Do events which are simultaneous in one inertial frame remain simultaneous when observed in another?

The answer is no.

Let us consider the two events shown in the spacetime diagram of Figure 6. The two inertial frames have been defined in standard configuration so that one of the events, Event 1, takes place at the spatial origin of both frames $(x = 0, x' = 0)$ at the instant $(ct = 0, ct' = 0)$ that the two frames coincide. (It is important to remember that the origin of the spacetime diagram is not the same thing as the origin of either inertial frame. The origin of any frame will trace out a *line* on any spacetime diagram that we draw.)

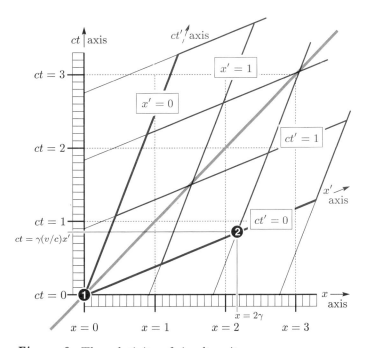

Figure 6   The relativity of simultaneity.

As far as an observer in the other inertial frame is concerned, Event 2 occurs at the same time as Event 1: $ct' = 0$. It simply occurs at a different place: $x' = 2$ instead of $x' = 0$. For her, the two events are simultaneous.

But what about for us? Inspection of the diagram shows immediately that for us, the two events take place not only at different places but also at different times. While the first event takes place at instant $ct = 0$, the second event takes place for us at a later time. (The Lorentz transformation equations can be used to show that it takes place at instant $ct = \gamma(v/c)\,x'$.)

So simultaneity is relative. It is relative to the particular inertial frame in which measurements are being made.

Not only that, but space and time are mixed together by the Lorentz transformations. Inspection of the diagram shows that the second event takes place for us not only at a later time than the first one, but also at a greater distance from the first one than it does in the other observer's inertial frame. For the other observer, the spatial distance between the two events is $\Delta x' = 2$. For us, the spatial distance between the two events is $\Delta x = 2\gamma$.

Time dilation

The experimentally observed phenomenon of time dilation is easy to understand with a spacetime diagram.

Suppose that two events take place for the other observer at the same place but at different times. An example would be two ticks of a clock which is at rest in her inertial frame of reference and therefore moving relative to our own. Two such events are shown in Figure 7. Remember that we still have $v/c = 0.4$.

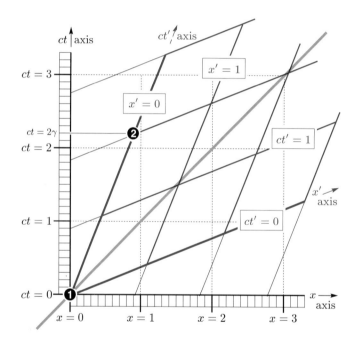

Figure 7 Time dilation.

In the other observer's inertial frame O', both events take place at position $x' = 0$. And the time interval $\Delta(ct')$ between the two events is two time units. But it can be seen immediately that in our own inertial frame, relative to which the clock is moving, the two events have a time interval between them of more than two time units: $\Delta(ct) = 2\gamma$.

Length contraction

In order to understand length contraction, two spacetime diagrams are needed rather than just one.

Let us consider a rigid rod which is at rest in the *other* observer's inertial frame. As time increases, the continuous existence of its two end-points traces out two parallel straight lines of events in our spacetime diagram, as shown in Figure 8.

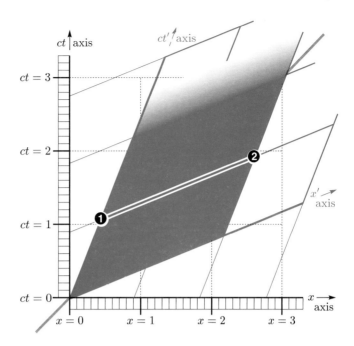

Figure 8 Length contraction: the rod at rest in the other observer's inertial frame.

If a snapshot is taken of the position of the two ends of the rod at the same instant in time in the other inertial frame, at say, $ct' = 1$ as shown, then the distance between the two events for the ends of the rod is simply $\Delta x' = 2$. The rod is two space-units long in the inertial frame in which it is at rest.

But now suppose that we take a snapshot of the rod in our own inertial frame, a frame in which the rod is moving.

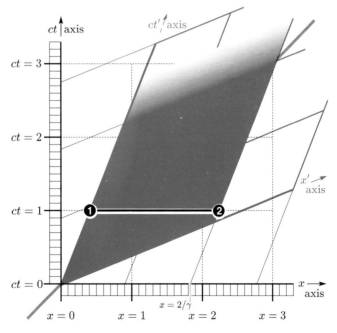

Figure 9 Length contraction: the rod moving in our inertial frame.

Again we look at where the two ends of the rod are, at the same instant in time. But the same instant in time means something different in our inertial frame, to what it means in the other inertial frame. If we take the snapshot in our own inertial frame at time $ct = 1$, then the position of the two ends is as shown in Figure 9. And their distance apart clearly is less than two units: $\Delta x = 2/\gamma$.

So the length of the moving rod as measured in our own inertial frame is contracted in the direction of the motion of the rod.

The travelling twin (and friends)

The phenomenon of the travelling twin who leaves Earth and comes back younger than his or her stay-at-home sibling is also easy to understand in terms of spacetime diagrams. This time it takes *three* diagrams.

We define the stay-at-home twin to be one of us. Then the first diagram is Figure 10. It shows how the travelling twin's path in spacetime looks to us, as we remain at rest in our own inertial frame. The travelling twin and friends begin at rest with us in our inertial frame O. Then they are accelerated into constant x-component of velocity $+v$ away from us at Event 1, and carry on travelling in inertial frame O' for two of our time units. Then they are decelerated and re-accelerated back towards us at Event 2, travelling at constant x-component of velocity $-v$ back towards us in inertial frame O'' for two more of our time units. Finally they are decelerated and brought to rest with us again at Event 3.

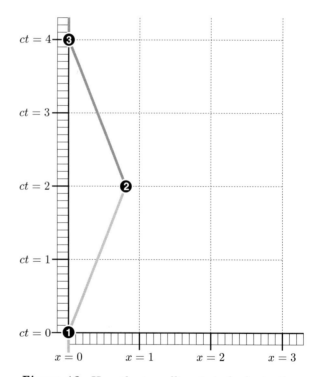

Figure 10   How the travelling twin looks to the stay-at-home twin.

We can represent their entire journey, as observed in our single inertial frame O, on one spacetime diagram.

But the picture is rather different for them. In order for us to represent what is observed in their various different inertial frames, we have to draw several different spacetime diagrams.

Firstly, we draw Figure 10 again. Before Event 1, the travellers are at rest with us in our inertial frame, and what they observe is the same as what we observe.

But after they undergo their first acceleration at Event 1, we have to draw a diagram like Figure 11 to represent what they observe in their new inertial frame O′. It can be seen by inspection that the time spent on their outward journey as observed in their own inertial frame O′ is less than it is as measured in our inertial frame O. We measure it as $\Delta(ct) = 2$. They measure it as $\Delta(ct') = 2/\gamma$.

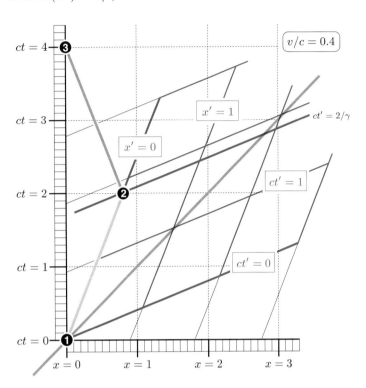

Figure 11   The travelling twin after the first acceleration.

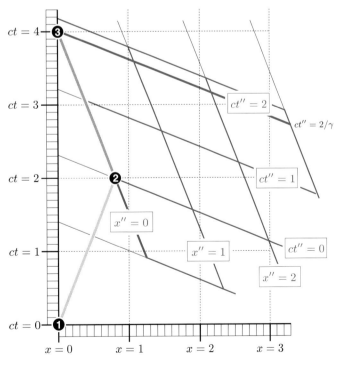

Figure 12   The travelling twin after the turn-around.

After they undergo their turn-around at Event 2, the diagram which we must draw for what they observe now is as in Figure 12. Now they are travelling at x-component of velocity $-v$ rather than $+v$ relative to us. Again, the time spent on their return voyage is less as measured in their inertial frame O'' than it is as measured in O. It can be seen by inspection that in O'', the elapsed time is $\Delta(ct'') = 2/\gamma$. In O it is again $\Delta(ct) = 2$.

The consequence is that the elapsed time for the journey as they measure it is less than the elapsed time as we measure it. We measure it as four time units in all. They measure it as four time units divided by the factor γ.

There is no paradox, because their situation and ours are not symmetrical. We sit still for the whole time in one inertial frame of reference, which means that we can draw their entire journey on one spacetime diagram. In contrast, they undergo three major accelerations and spend their time in a total of four successive inertial frames (counting their start at rest with us in the same inertial frame as our own). In order to draw our apparent motion relative to them, they have to draw four different spacetime diagrams — one for each of the inertial frames in which they find themselves — and these diagrams cannot be superimposed.

So they cannot draw a spacetime diagram from which any 'elapsed time' of our apparent journey relative to them can be read off. The only diagram from which an elapsed time for the whole journey can be determined is ours, and it says that they age less than we do.

Lest there be any further doubt: this result has been verified repeatedly by experiment.

7 Essay exercises for Blocks 1 and 2

We suggest that, as in an exam, you read through all the essay topics and choose one. Of course, you may then wish to try others, or at least study the model answers. Write as briefly as is consistent with clarity. Remember that marks are awarded for relevant points, not waffle! These five essays are somewhat shorter than the ones set in the examination, so allow yourself 30–40 minutes for each, and aim for a total length of 350–600 words for each. Note the appendix on essay writing

1 Write an essay explaining the relevance of the homogeneity and isotropy of space to the following topics in Newtonian mechanics:

(i) The process of setting up a Cartesian coordinate system.

(ii) Models of the Solar System from Ptolemy to Newton.

(iii) The conservation laws of momentum and angular momentum.

2 Write an essay on Newton's first and second laws of motion, including a discussion of the points listed below:

(i) Does the first law do more than define an inertial frame?

(ii) Does the second law do more than define force?

(iii) What would be required for either law to be disproved experimentally?

3 Write an essay on the principle of relativity in Newtonian mechanics. You should include a clear statement of the principle, a definition of an inertial frame, and an explanation of the relevance of the homogeneity of space and time and the isotropy of space.

Briefly indicate the logical flaws in the objections listed below:

(i) Gravity is weaker on the Moon than on the Earth. This disproves homogeneity of space.

(ii) The weather changes from season to season. This disproves homogeneity of time.

(iii) Objects fall downwards, not upwards. This disproves isotropy of space.

4 'In the new picture, the length of a moving rod is less than that of a stationary rod, and the time intervals between ticks on a moving clock are longer than those on a stationary one.'

Write an essay explaining the meaning of this quotation from the Introduction to Block 2. Your essay should make clear what is to be understood by *observing*, and how this differs from *seeing*.

5 Write an essay comparing the world-view of Einstein's special theory of relativity with that of Newton and Galileo. Indicate to what extent they are in agreement, and explain in what ways they differ in relation to the following points:

(i) the postulates A1 to A10 in Block 1;

(ii) the principle of relativity;

(iii) the conservation laws of momentum, mass and energy.

8 Solutions to consolidation exercises

8.1 Exercises for Units 1–3

Exercise 1   This exercise is based on Equations 1, 3, 4 and 5 from Unit 1.

(i) From Equations 3, 4 and 5,

$$x_A^1 = 5 \times \sin 112° \times \cos 37° = 3.702$$
$$x_A^2 = 5 \times \sin 112° \times \sin 37° = 2.790$$
$$x_A^3 = 5 \times \cos 112° = -1.873,$$

so that $\mathbf{x}_A = (3.702, 2.790, -1.873)$.

(ii) Using Equation 1,

$$r_B = |\mathbf{x}_B| = \sqrt{2^2 + 3^2 + 7^2} = 7.874.$$

Now from Equation 5,

$$\cos\theta_B = \frac{7}{r_B} = 0.889,$$

which gives $\theta_B = 27°$. This is the only possible solution as the range of the polar angle θ is zero to 180°.

From Equations 3 and 4,

$$\cos\phi_B = \frac{2}{r_B \sin\theta_B} = 0.555$$
$$\sin\phi_B = \frac{-3}{r_B \sin\theta_B} = -0.832.$$

As the range of the azimuthal angle ϕ is zero to 360°, both the sine and cosine are needed to determine the angle. In the present case, with the cosine positive but the sine negative, the angle is in the fourth quadrant, and is therefore 304°. Thus the spherical polar coordinates for $\mathbf{x}_B$ are; $r_B = 7.874$, $\theta_B = 27°$ and $\phi_B = 304°$.

Exercise 2   Coordinates: x^1(east), x^2(north). (x^3 not applicable.) The overall displacement is the sum of the displacement vectors for the three straight sections of the course followed by the ship, i.e.

$$\overrightarrow{AD} = \overrightarrow{AB} + \overrightarrow{BC} + \overrightarrow{CD}$$
$$= (0, 80, 0) + (100/\sqrt{2}, 100/\sqrt{2}, 0) + (-50, 0, 0)$$
$$= (20.7, 150.7, 0) \text{ nautical miles.}$$

The distance from A to D is given by the magnitude of the displacement vector $\overrightarrow{AD}$, which is

$$|\overrightarrow{AD}| = \sqrt{20.7^2 + 150.7^2} = 152.1 \text{ nautical miles.}$$

The vector $\overrightarrow{AD}$ makes an angle $\arctan(20.7/150.7) = 7.8°$ with north, and has a positive easterly component. It follows that D is on a bearing north 7.8° east from A.

The total distance travelled is 230 nautical miles and at 22 knots, this takes 10.45 hours. The average velocity vector of the ship, from A to D is thus $152.1/10.45 = 14.6$ knots on a bearing north 7.8° east.

Exercise 3   Denoting the origin by O, we have

$$\overrightarrow{AB} = \overrightarrow{OB} - \overrightarrow{OA} = (0, -2, -2)$$
$$\overrightarrow{BC} = \overrightarrow{OC} - \overrightarrow{OB} = (-2, 2, 0)$$
$$\overrightarrow{CA} = \overrightarrow{OA} - \overrightarrow{OC} = (2, 0, 2).$$

The lengths of all these vectors are $\sqrt{2^2 + 2^2} = 2\sqrt{2}$, so ABC is equilateral and all its internal angles must be 60°. We can check this by calculating the scalar product of any pair of sides, e.g. $\overrightarrow{AB} \cdot \overrightarrow{BC}$, and obtain

$$(0, -2, -2) \cdot (-2, 2, 0) = -2 \times 2 = -4.$$

As the product of their lengths is 8, the cosine of the angle between these two vectors is -0.5, (see Equation 26 of Unit 1) and the angle is therefore 120°. However, this refers to the external angle of the triangle, so there is no discrepancy. (See Figure 13.)

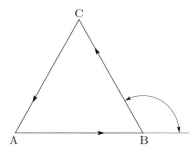

Figure 13

Exercise 4   (i) If latitude is defined as positive in the northern hemisphere, and negative in the southern, the co-latitude, θ, covers the range zero to 180°; e.g. for London $\theta_L = 90° - 51° = 39°$, and for Montevideo $\theta_M = 90° - (-35°) = 125°$. The angle ϕ may be defined as the longitude if towards the east, and as $(360° -$ longitude$)$ if towards the west. Thus for London $\phi_L = 0$, and for Montevideo $\phi_M = 360° - 56° = 304°$. In this way, polar coordinates (R, θ_P, ϕ_P) can be given for any point, P, on the Earth. The Cartesian components of $\overrightarrow{OP}$ are given by Equations 3, 4 and 5 from Unit 1. Clearly, $|\overrightarrow{OP}| = R$.

(ii) For any two points P_1 and P_2 the angle Ω, between the vectors $\overrightarrow{OP_1}$ and $\overrightarrow{OP_2}$ is given in terms of their scalar product by

$$\cos\Omega = (\overrightarrow{OP_1} \cdot \overrightarrow{OP_2})/R^2 = (x_1^1 x_2^1 + x_1^2 x_2^2 + x_1^3 x_2^3)/R^2$$
$$= \sin\theta_1 \sin\theta_2(\cos\phi_1 \cos\phi_2 + \sin\phi_1 \sin\phi_2)$$
$$\quad + \cos\theta_1 \cos\theta_2$$
$$= \sin\theta_1 \sin\theta_2 \cos(\phi_1 - \phi_2) + \cos\theta_1 \cos\theta_2.$$

The length of the arc between P_1 and P_2, the great circle route, is thus $R\Omega$ with Ω in radians. Alternatively, it is given in nautical miles by Ω expressed in minutes of arc.

Putting in the appropriate values for London and Montevideo gives $\cos \Omega = -0.1575$, from which $\Omega = 99°$. The distance by the great circle route is therefore $99 \times 60 = 5940$ nautical miles.

Exercise 5 (i) The satellite is positioned above the Equator at longitude $60°$ E, so that $\theta_S = 90°$ and $\phi_S = 60°$. From Equations 3, 4 and 5 of Unit 1, its position vector in the Cartesian system is thus

$$\overrightarrow{OS} = R(6.626 \times \cos 60°, 6.626 \times \sin 60°, 0)$$
$$= R(3.313, 5.7383, 0).$$

(ii) For London, we have (cf. Exercise 4)

$$\overrightarrow{OL} = R(\sin 39°, 0, \cos 39°)$$
$$= R(0.6293, 0, 0.7771).$$

The displacement vector from London to the satellite is

$$\overrightarrow{LS} = \overrightarrow{OS} - \overrightarrow{OL} = R(2.6837, 5.7383, -0.7771),$$

and the distance from London to the satellite is thus

$$|\overrightarrow{LS}| = (\sqrt{2.6837^2 + 5.7383^2 + 0.7771^2})R = 6.3823R.$$

(iii) The cosine of the angle between $\overrightarrow{LS}$ and $\overrightarrow{OL}$ is

$$\overrightarrow{LS} \cdot \overrightarrow{OL} / |\overrightarrow{LS}||\overrightarrow{OL}|$$
$$= (2.6837 \times 0.6293 - 0.7771^2)/6.3823 = 0.17.$$

This means that $\overrightarrow{LS}$ makes an angle $80.2°$ with the vertical through London, and an angle $90° - 80.2° = 9.8°$ with the horizontal. A dish in London should therefore be pointed at an elevation of $9.8°$ above the horizon to receive signals from the satellite.

Exercise 6 Using the same Cartesian axes as in Exercise 2, and placing the submarine at the origin of coordinates the position vector (in nautical miles) of the warship at time t hours is given by

$$x_W^1 = 22 \sin 45° \, t = 11\sqrt{2}t$$
$$x_W^2 = 2.5 + 22 \cos 45° \, t = 2.5 + 11\sqrt{2}t.$$

Similarly, if the torpedo is fired on a course north ϕ degrees east, its position is

$$x_T^1 = 55t \sin \phi$$
$$x_T^2 = 55t \cos \phi.$$

The ship and the torpedo collide if their position vectors are equal at some instant (or if their world-lines intersect), and this requires $x_T^1 = x_W^1$ and $x_T^2 = x_W^2$. From the first of these we find

$$\phi = \arcsin(\sqrt{2}/5) = 16.43°.$$

Substituting this value of ϕ into the second equation gives $t = 6.72 \times 10^{-2}$ hours. The torpedo was therefore fired on a course north $16.43°$ east, and reached its target in just over 4 minutes.

Exercise 7 (i) The velocity vector is obtained directly by differentiating with respect to time, giving

$$\mathbf{v} = d\mathbf{x}/dt = \omega(-3 \cos \omega t, 4 \cos \omega t, -5 \sin \omega t),$$

and a second differentiation gives the acceleration vector as

$$\mathbf{a} = d\mathbf{v}/dt = \omega^2(3 \sin \omega t, -4 \sin \omega t, -5 \cos \omega t).$$

(ii) By comparing $\mathbf{a}$ with $\mathbf{x}$, we can see immediately that $\mathbf{a} = -\omega^2 \mathbf{x}$, i.e. the acceleration is proportional to minus the displacement from the origin. Thus $\mathbf{a}$ and $\mathbf{x}$ are parallel (more precisely, antiparallel) and it follows that $\mathbf{x} \times \mathbf{a} = \mathbf{0}$. Of course, this could have been verified by calculating the vector product directly.

For $\mathbf{x} \times \mathbf{v}$, we have the components

$$x^2 v^3 - x^3 v^2 = -20\omega \sin^2 \omega t - 20\omega \cos^2 \omega t = -20\omega$$
$$x^3 v^1 - x^1 v^3 = -15\omega \cos^2 \omega t - 15\omega \sin^2 \omega t = -15\omega$$
$$x^1 v^2 - x^2 v^1 = -12\omega \sin \omega t \cos \omega t + 12\omega \sin \omega t \cos \omega t = 0.$$

This vector is constant in time, as expected. The reason for this is readily appreciated by noting that

$$\frac{d}{dt}(\mathbf{x} \times \mathbf{v}) = (\frac{d\mathbf{x}}{dt} \times \mathbf{v}) + (\mathbf{x} \times \frac{d\mathbf{v}}{dt})$$
$$= (\mathbf{v} \times \mathbf{v}) + (\mathbf{x} \times \mathbf{a})$$
$$= \mathbf{x} \times \mathbf{a} = \mathbf{0}.$$

(Remember $\mathbf{v} \times \mathbf{v} = \mathbf{0}$.)

(iii) We have

$$|\mathbf{x}|^2 = \mathbf{x} \cdot \mathbf{x} = (x^1)^2 + (x^2)^2 + (x^3)^2$$
$$= 9 \sin^2 \omega t + 16 \sin^2 \omega t + 25 \cos^2 \omega t = 25,$$

which implies that the particle maintains a constant distance of 5 units from the origin throughout the motion. It follows from this that

$$\frac{d|\mathbf{x}|^2}{dt} = 2\mathbf{x} \cdot \mathbf{v} = 0,$$

which means that the velocity is always at right angles to the displacement from the origin. The speed of the particle is also constant, since

$$|\mathbf{v}|^2 = \mathbf{v} \cdot \mathbf{v} = 25\omega^2.$$

We deduce from this that the particle is moving, with constant speed, on the surface of a sphere centred at the origin. In fact, we can say a little more. We know that the acceleration vector, $\mathbf{a}$, is parallel to $\mathbf{x}$, and therefore has no component parallel to the surface of the sphere. This means that the particle cannot change direction on the surface, and therefore moves on a great circle. The particle is thus moving in a circle of radius 5 units at a speed 5ω, or angular speed ω.

(iv) We saw in part (ii) that

$$\mathbf{x} \times \mathbf{v} = (-20\omega, -15\omega, 0) = 25\omega(-0.8, -0.6, 0).$$

As this *constant* vector is perpendicular to both $\mathbf{x}$ and $\mathbf{v}$, the motion takes place in a fixed plane perpendicular to the unit vector $\hat{\mathbf{n}} = (-0.8, -0.6, 0)$, which lies in the **1,2**-plane making angles of $37°$ and $53°$, respectively with the negative **1**- and **2**-axes. The vector $\hat{\mathbf{n}}$ is, effectively, the axle of the 'wheel'. If the thumb of a right hand is aligned with $\hat{\mathbf{n}}$, the direction in which the fingers curl indicates the sense of a *positive* rotation, i.e. anticlockwise when viewed from the direction in which the thumb is pointing.

Exercise 8   (i) Differentiation with respect to time gives

$$\mathbf{v} = d\mathbf{x}/dt = R\,\omega(1 - \cos\omega t, \sin\omega t, 0)$$

and

$$\mathbf{a} = d\mathbf{v}/dt = R\,\omega^2(\sin\omega t, \cos\omega t, 0).$$

(ii) The particle is instantaneously at rest whenever $\cos\omega t = 1$ and $\sin\omega t = 0$, i.e. when $\omega t = 2\pi n$, where n is an integer. Substitution of this value of t into the formulae for $\mathbf{x}$, $\mathbf{v}$ and $\mathbf{a}$ gives

$$\mathbf{x} = (2\pi n R, 0, 0),$$
$$\mathbf{v} = (0, 0, 0)$$
$$\mathbf{a} = (0, R\,\omega^2, 0).$$

(iii) The path, known as a *cycloid*, is that of the valve on a bicycle inner-tube (strictly a stone wedged in the tread of the tyre) when the machine is being ridden along a level straight road. This can be seen by writing the position, $\mathbf{x}$, as the vector sum of two vectors, i.e.

$$\mathbf{x} = (R\omega t, R, 0) + (-R\sin\omega t, -R\cos\omega t, 0).$$

The first vector represents the steady forward motion of the hub of the wheel at speed $R\omega$ and height R above the surface of the road. The second vector describes the uniform circular motion of the valve about the hub. Clearly R is the radius of the wheel.

The path is easily sketched using the calculated values in the table below. It is illustrated in Figure 14.

ωt	x^1	x^2
0	0	0
$\pi/2$	$(\pi/2 - 1)R$	R
π	πR	$2R$
$3\pi/2$	$(3\pi/2 + 1)R$	R
2π	$2\pi R$	0
$5\pi/2$	$(5\pi/2 - 1)R$	R
3π	$3\pi R$	$2R$
$7\pi/2$	$(7\pi/2 + 1)R$	R
4π	$4\pi R$	0

Exercise 9   (i) Let S′ be the new reference frame which results from rotating S through π radians about the **3**-axis. Then, referring to Figure 15(a), we see that the **1′**-axis coincides with the negative **1**-axis, and the **2′**-axis coincides with the negative **2**-axis. The **3**- and **3′**-axes coincide. This situation is summarized by the relation

$$(x^1, x^2, x^3)' = (-x^1, -x^2, x^3)$$

which connects the Cartesian coordinates of the *same point* with respect to the two frames S and S′. In a similar way, let S″ be the new frame which results from a reflection in the **1′**, **2′**-plane. The **3″**-axis now coincides with the negative **3′**-axis, while the **1″**- and **2″**-axes coincide with the **1′**- and **2′**-axes respectively. We can relate the coordinates of a point in S″ to those of the same point in S′, and hence to the coordinates of the same point in S:

$$(x^1, x^2, x^3)'' = (x^1, x^2, -x^3)' = (-x^1, -x^2, -x^3).$$

Thus *all* the axes of S″ coincide with the corresponding negative axes in S. Note that S″ is a left-handed rather than a conventional right-handed frame.

(ii) In this case, the frame S′ results from a positive rotation of S about the **3**-axis by $\pi/2$ radians. This is shown in Figure 15(b). The sense of the rotation is positive according to the right-hand rule. Coordinates in the two frames are clearly related by

$$(x^1, x^2, x^3)' = (x^2, -x^1, x^3).$$

The frame S″ is obtained by rotating S′ through π radians about the **1′**-axis, as shown in Figure 15(c). Coordinates in S′ and S″ are related by

$$(x^1, x^2, x^3)'' = (x^1, -x^2, -x^3)'.$$

Combining the previous two relations, we obtain

$$(x^1, x^2, x^3)'' = (x^2, x^1, -x^3).$$

As the axes have undergone rotations only in this case, i.e. no reflections, S″ is a right-handed frame.

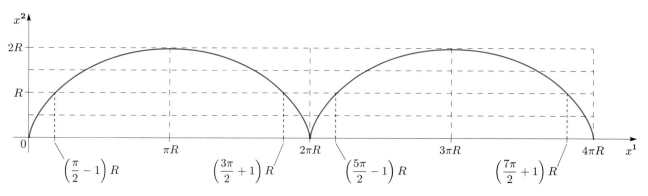

Figure 14

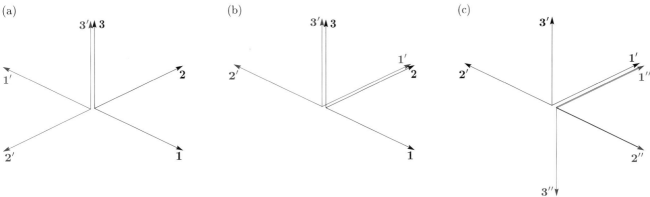

Figure 15

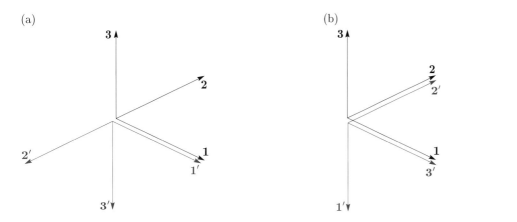

Figure 16

Exercise 10   (i) (a) Referring to Figure 16(a), it can be seen that under this rotation $(x^1, x^2, x^3)' = (x^1, -x^2, -x^3)$, so that in S'

$$\mathbf{x} = (1, 2, -4) \qquad \mathbf{v} = (-5, -2, 0).$$

(b) Coordinates transform according to $(x^1, x^2, x^3)' = (-x^3, x^2, x^1)$ under this rotation, as shown in Figure 16(b). Thus, in S'

$$\mathbf{x} = (-4, -2, 1) \qquad \mathbf{v} = (0, 2, -5).$$

(c) Here, S' is a left-handed frame having its **3**-axis reversed compared to S. Thus $(x^1, x^2, x^3)' = (x^1, x^2, -x^3)$, and in S'

$$\mathbf{x} = (1, -2, -4) \qquad \mathbf{v} = (-5, 2, 0).$$

(d) For any position vector, $(x^1, x^2, x^3)' = (x^1 - 3, x^2 - 3, x^3 - 1)$. As a velocity vector is unchanged by any static translation, we have in S',

$$\mathbf{x} = (-2, -5, 3) \qquad \mathbf{v} = (-5, 2, 0).$$

(ii) In S, the scalar product is given by

$$\mathbf{x} \cdot \mathbf{v} = (1 \times -5) + (-2 \times 2) = -9.$$

Transformations (a) and (b) are rotations, so that the scalar product is necessarily invariant under these. In transformation (c), the **3**-axis is reversed. This means that the **3**-components of both the vectors change sign, and the scalar product is therefore unaffected. Thus $\mathbf{x} \cdot \mathbf{v} = -9$ in the S' frames generated by all the transformations (a)–(c).

Exercise 11   Parts (i) and (ii) depend on the units of quantities derived from position, velocity and acceleration, which have units of m, $\mathrm{m\,s^{-1}}$ and $\mathrm{m\,s^{-2}}$ respectively.

(i) A. This has $\mathrm{s^{-2}}$ in the numerator and also in the denominator. It therefore does not depend on the unit of time. All the remaining options depend on the unit of time.

(ii) B. This has $\mathrm{m^2}$ in both numerator and denominator, and does not depend on the unit of length. All other options depend on the unit of length.

(iii) A and D. These options are functions of velocity and acceleration only, and are therefore unaffected by static translations of the frame of reference. All the other options depend on position, which changes under a translation.

(iv) A, B and E. These quantities are all scalars, and therefore invariant under all static rotations of the reference frame. The other two options are both vectors, and have different components according to the orientation of the axes.

Exercise 12   (i) From Figure 1,

$$x^1 = r_1 \cos\theta \qquad x^2 = r_2 \sin\theta,$$

so that $\cos\theta = x^1/r_1$ and $\sin\theta = x^2/r_2$. Thus, substituting these into the identity

$$\sin^2\theta + \cos^2\theta = \left(\frac{x^1}{r_1}\right)^2 + \left(\frac{x^2}{r_2}\right)^2 = 1.$$

(ii) Referring to Figure 1 and applying Pythagoras' theorem, we obtain

$$|F_1P|^2 = (|F_1O| + x^1)^2 + (x^2)^2$$
$$= r_1^2(\varepsilon + \cos\theta)^2 + r_1^2(1 - \varepsilon^2)\sin^2\theta$$
$$= r_1^2(1 + \varepsilon\cos\theta)^2,$$

so that $|F_1P| = r_1(1 + \varepsilon\cos\theta)$.

Similarly

$$|F_2P|^2 = (-|F_2O| + x^1)^2 + (x^2)^2$$
$$= r_1^2(-\varepsilon + \cos\theta)^2 + r_1^2(1 - \varepsilon^2)\sin^2\theta$$
$$= r_1^2(1 - \varepsilon\cos\theta)^2,$$

and hence $|F_2P| = r_1(1 - \varepsilon\cos\theta)$.

It follows immediately that $|F_1P| + |F_2P| = 2r_1$.

(iii) From Figure 1, if $r = |F_1P|$ and $\phi = \angle OF_1P$,

$$r\cos\phi = |F_1O| + x^1 = r_1(\varepsilon + \cos\theta).$$

But from part (ii), we have

$$r = r_1(1 + \varepsilon\cos\theta).$$

Eliminating $\cos\theta$ between these two equations gives the polar equation of the ellipse, referred to F_1 as origin,

$$r(1 - \varepsilon\cos\phi) = r_1(1 - \varepsilon^2).$$

In a similar way, if $r = |F_2P|$ and $\phi = \angle OF_2P$,

$$r\cos\phi = |F_2O| - x^1 = r_1(\varepsilon - \cos\theta).$$

Again, using a previous result, $|F_2P| = r_1(1 - \varepsilon\cos\theta)$, and elimination of $\cos\theta$ leads to the same polar equation, this time referred to F_2 as origin.

Exercise 13 (i) From Figure 1 of this unit, or Figure 6 in Unit 2, the position of the particle is given by (since r_1 and r_2 are constants)

$$\mathbf{x} = (r_1\cos\theta, r_2\sin\theta, 0)$$

and its velocity vector is therefore

$$\mathbf{v} = (-r_1\sin\theta, r_2\cos\theta, 0)\frac{d\theta}{dt}.$$

(ii) The rate of sweeping out of area is given in Section 2.2 Unit 2 as

$$\tfrac{1}{2}|\mathbf{x} \times \mathbf{v}|,$$

and this is known to be a constant. As the total area of the ellipse is $\pi r_1 r_2$, this constant value must be $\pi r_1 r_2/T$. The only non-zero component of $\mathbf{x} \times \mathbf{v}$ is the **3**-component, given by

$$x^1v^2 - x^2v^1 = r_1r_2\frac{d\theta}{dt}.$$

Thus

$$\tfrac{1}{2}|\mathbf{x} \times \mathbf{v}| = \tfrac{1}{2}r_1r_2\frac{d\theta}{dt} = \pi r_1r_2/T,$$

from which it follows that

$$\frac{d\theta}{dt} = \frac{2\pi}{T}.$$

(iii) The acceleration vector is now seen to be

$$\mathbf{a} = (-r_1\cos\theta, -r_2\sin\theta, 0)(2\pi/T)^2 = -(2\pi/T)^2\mathbf{x},$$

and its magnitude is

$$|\mathbf{a}| = (2\pi/T)^2|\mathbf{x}|.$$

It is directed from the particle towards the centre of the ellipse.

The motion described in this exercise is exemplified by the small amplitude oscillations of the bob of a simple pendulum when, instead of oscillating back and forth along a straight line as is customary, it follows a closed path in *two dimensions*. This arrangement is often called the 'spherical pendulum'. The one-dimensional motion of a simple pendulum is treated in some detail in SAQs 8–12 in Unit 2. The horizontal force on the bob, which is provided by the horizontal component of the tension in the supporting string, is proportional to (minus) the displacement vector of the bob from the origin (i.e. its equilibrium position). For this reason, the angular momentum is a conserved quantity, and Kepler's second law is obeyed by the system. Provided the oscillations are of an amplitude which is small compared to the length of the pendulum, the oscillation period, T, is independent of the amplitude. For a pendulum of length l,

$$T = 2\pi\sqrt{l/g},$$

where g is the acceleration due to gravity. Uniform circular motion (Section 4.5 Unit 1) is a special case of the motion.

Exercise 14 (i) We assume that the satellite moves in a circular orbit of radius R. (The orbital radius must be *slightly* greater than R, but we may neglect the difference.) The satellite must obey Kepler's second law, and therefore moves with uniform angular speed $\omega = 2\pi/T$, where T is its orbital period. It is shown in Section 4.5 Unit 1 that a particle moving uniformly in a circle suffers an acceleration towards the centre (sometimes called the *centripetal* acceleration) of magnitude ω^2R. For a satellite of mass m, the force sustaining this acceleration is $m\omega^2R$ and must be equal to the weight, mg. The mass cancels out and we are left with

$$\omega^2R = (2\pi/T)^2 R = g,$$

so that $T = 2\pi\sqrt{R/g}$ as required. Using the values given, we find

$$T = 2\pi\sqrt{6.37 \times 10^6/9.8} = 5066\,\text{s}.$$

Thus the period of the hypothetical satellite is 1.407 hours.

(ii) For a geostationary satellite, the orbit should be circular, and of period 24 hours. The radius, R_G, of the orbit is given immediately by Kepler's third law;

$$R_G = (24/1.407)^{2/3}R = 6.626R = 4.22 \times 10^7\,\text{m}.$$

Exercise 15 An object of mass m undergoing uniform circular motion of radius r and angular speed ω, accelerates towards the centre of rotation, and the acceleration is sustained by a force of magnitude $m\omega^2 r$. (See Section 4.3 Unit 2.) The ratio, f, of this force to the weight, mg, of the object is $f = \omega^2 r/g$. For the astronaut in the question, $f = 9$, so that

$$\omega = \sqrt{gf/r} = \sqrt{9.8 \times 9/5} = 4.2 \text{ radian s}^{-1}.$$

Exercise 16 (i) The frame S′ is obtained from S by interchanging the **1**- and **2**-directions. The relation between coordinates in the two frames is therefore

$$(x^{\mathbf{1}}, x^{\mathbf{2}}, x^{\mathbf{3}})' = (x^{\mathbf{2}}, x^{\mathbf{1}}, x^{\mathbf{3}}),$$

and this is applicable to the components of any vector. Thus at $t = t_1$, the positions and velocities of A and B referred to S′ are

$$\mathbf{x}_A = (0, -X, 0) \qquad \mathbf{v}_A = (0, V, 0)$$

and

$$\mathbf{x}_B = (-X, 0, 0) \qquad \mathbf{v}_B = (V, 0, 0).$$

If O′ now regards A as B and B as A, which is possible because the billiard balls are identical, her description of the initial configuration evidently agrees with that of O.

(ii) The principle of relativity implies that the agreement between O and O′ must persist throughout the subsequent motion. It follows that when, at $t = t_2$, O observes ball A to have position and velocity

$$\mathbf{x}_A = (a, b, 0) \qquad \mathbf{v}_A = (u, v, 0),$$

O′ also observes ball A to have the same position and velocity. However, she attributes these parameters to what O regards as ball B, and consequently *its* position and velocity in S must be given by

$$\mathbf{x}_B = (b, a, 0) \qquad \mathbf{v}_B = (v, u, 0).$$

(iii) If we equate the total momentum in either the **1**- or **2**-directions in S at $t = t_1$ with its value at $t = t_2$, we find (dividing by the common mass)

$$u + v = V.$$

By assumption, the total kinetic energy of the system is the same at $t = t_1$ and $t = t_2$, and it follows that

$$u^2 + v^2 = V^2.$$

These two relations between u and v imply that $uv = 0$, and hence that

$$u = 0, \quad v = V.$$

(The alternative solution, namely $u = V$, $v = 0$, can be disregarded on the grounds that it requires the motion of neither ball to be affected by the collision; they would pass through each other without interaction!)

(iv) The **3**-component of angular momentum about the origin at $t = t_1$ is clearly zero, as the initial paths of both balls pass through the origin. At $t = t_2$,

$$\mathbf{x}_A \times \mathbf{v}_A = (a, b, 0) \times (0, V, 0) = (0, 0, aV)$$

$$\mathbf{x}_B \times \mathbf{v}_B = (b, a, 0) \times (V, 0, 0) = (0, 0, -aV),$$

so that total angular momentum is again zero, whatever the values of a and b. Thus conservation laws cannot give us a and b.

[A detailed analysis of the motion of the balls before and after the collision shows that $a = -\sqrt{2}R$, where R is the radius of the balls. The value of b is governed by the time elapsed between the collision and $t = t_2$. If the collision occurs at $t = (t_1 + t_2)/2$, then $b = X + a$.]

Exercise 17 (i) If the coordinates $\mathbf{x}_1$ and $\mathbf{x}_2$ of the two particles are interchanged, $\mathbf{x}$ changes sign and

$$\mathbf{F}_{21} = -f(|\mathbf{x}|)\frac{\mathbf{x}}{|\mathbf{x}|} = -\mathbf{F}_{12},$$

i.e. the forces between the particles are equal and opposite. It is clear also that they are parallel to the separation vector, $\mathbf{x}$. This establishes N3 for the system (cf. Section 5.5 Unit 3).

(ii) Adding the two equations given leads directly to

$$\frac{\mathrm{d}^2(m_1\mathbf{x}_1 + m_2\mathbf{x}_2)}{\mathrm{d}t^2} = M\frac{\mathrm{d}^2\mathbf{X}}{\mathrm{d}t^2} = \mathbf{0}.$$

This means that the acceleration of the centre of mass is zero, and its velocity is therefore constant.

(iii) In the new coordinate system, $\mathbf{X} = \mathbf{0}$ so that

$$m_1\mathbf{x}_1 + m_2\mathbf{x}_2 = \mathbf{0}.$$

Recalling from part (i) that $\mathbf{x}_1 - \mathbf{x}_2 = \mathbf{x}$, we find immediately

$$\mathbf{x}_1 = \frac{m_2}{M}\mathbf{x} \qquad \mathbf{x}_2 = -\frac{m_1}{M}\mathbf{x}.$$

Substitution of these expressions into the two equations of motion gives, in both cases,

$$\mu\frac{\mathrm{d}^2\mathbf{x}}{\mathrm{d}t^2} = f(|\mathbf{x}|)\frac{\mathbf{x}}{|\mathbf{x}|}.$$

The new reference frame is necessarily inertial because the centre of mass of the system has zero acceleration in the original inertial frame.

This new equation of motion may be interpreted as follows. The position vectors of the two particles, relative to their centre of mass, are both proportional to their separation vector, $\mathbf{x}$. Therefore both particles move around the centre of mass on paths which are just scaled reproductions of their *relative* motion; this being equivalent to the motion of a *single* particle of mass μ about a fixed centre of attraction.

(iv) It is easily seen from the definition of μ that $m_1 m_2 = \mu M$, so that in the case of a gravitational interaction, for which

$$f(|\mathbf{x}|) = -\frac{Gm_1 m_2}{|\mathbf{x}|^2},$$

the reduced mass, μ, cancels out of the equation of motion, which then reduces to

$$\frac{\mathrm{d}^2\mathbf{x}}{\mathrm{d}t^2} = \frac{-GM}{|\mathbf{x}|^3}\mathbf{x}.$$

It follows that the relative motion of the stars depends only on their total mass, M. In particular,

Kepler's third law takes the form

$$\frac{r_1^3}{T^2} = \frac{GM}{(2\pi)^2},$$

and measurements of T and r_1 for the relative motion provide a value of M only. The masses of the individual stars can be obtained if the position of the centre of mass can be located. The ratio of the distances from the centre of mass to the stars then provides the ratio m_2/m_1.

Exercise 18 (i) When i and j are interchanged, $\mathbf{x}_{ij}$ changes sign, but $|\mathbf{x}_{ij}|$ is unaltered. It follows immediately from this that $\mathbf{F}_{ij} = -\mathbf{F}_{ji}$. Moreover, these forces are clearly parallel to the line of centres of the particles. These two conditions establish N3, as stated in Section 5.5 Unit 3, for this system.

(ii) (a) $\mathbf{F}_{ij} = \dfrac{Q_i Q_j \mathbf{x}_{ij}}{4\pi\varepsilon_0 |\mathbf{x}_{ij}|^3}$.

(b) $\mathbf{F}_{ij} = -k\mathbf{x}_{ij}$.

Exercise 19 (i) Expressions for $\mathbf{P}$ and $\mathbf{J}$ in a two-particle system are given in Section 5.6 Unit 3. Extending these in the obvious way to the case of three particles we have

$$\mathbf{P} = m_1\mathbf{v}_1 + m_2\mathbf{v}_2 + m_3\mathbf{v}_3,$$

and

$$\mathbf{J} = m_1\mathbf{x}_1 \times \mathbf{v}_1 + m_2\mathbf{x}_2 \times \mathbf{v}_2 + m_3\mathbf{x}_3 \times \mathbf{v}_3.$$

(ii) To show that $\mathbf{P}$ is constant, consider

$$d\mathbf{P}/dt = m_1\mathbf{a}_1 + m_2\mathbf{a}_2 + m_3\mathbf{a}_3.$$

Now $m_1\mathbf{a}_1$ is equal to the *total* force acting on particle 1. The other two particles each exert a force on this particle, so that

$$m_1\mathbf{a}_1 = \mathbf{F}_{12} + \mathbf{F}_{13},$$

and similar expressions give the total forces on the other two particles. Thus, from N3

$$d\mathbf{P}/dt = \mathbf{F}_{12} + \mathbf{F}_{13} + \mathbf{F}_{21} + \mathbf{F}_{23} + \mathbf{F}_{31} + \mathbf{F}_{32} = \mathbf{0}.$$

In a similar way we consider

$$d\mathbf{J}/dt = m_1\mathbf{x}_1 \times \mathbf{a}_1 + m_2\mathbf{x}_2 \times \mathbf{a}_2 + m_3\mathbf{x}_3 \times \mathbf{a}_3.$$

(Terms of the form $m_1\mathbf{v}_1 \times \mathbf{v}_1$ which arise in the differentiation are all zero.) Substituting for $m_1\mathbf{a}_1$ etc., as before, we obtain

$$d\mathbf{J}/dt = \mathbf{x}_1 \times (\mathbf{F}_{12} + \mathbf{F}_{13})$$
$$+ \mathbf{x}_2 \times (\mathbf{F}_{21} + \mathbf{F}_{23}) + \mathbf{x}_3 \times (\mathbf{F}_{31} + \mathbf{F}_{32}).$$

By again making use of N3, we can write this as

$$d\mathbf{J}/dt = (\mathbf{x}_1 - \mathbf{x}_2) \times \mathbf{F}_{12}$$
$$+ (\mathbf{x}_1 - \mathbf{x}_3) \times \mathbf{F}_{13} + (\mathbf{x}_2 - \mathbf{x}_3) \times \mathbf{F}_{23}$$
$$= \mathbf{x}_{12} \times \mathbf{F}_{12} + \mathbf{x}_{13} \times \mathbf{F}_{13} + \mathbf{x}_{23} \times \mathbf{F}_{23}.$$

Since in each case the force acts along the line of separation, each of these terms is zero, so $d\mathbf{J}/dt = \mathbf{0}$. This completes the proof that both $\mathbf{P}$ and $\mathbf{J}$ are constant.

Exercise 20 (i) Each *pair* of particles contributes to the total potential energy, which is therefore

$$U = U(|\mathbf{x}_{12}|) + U(|\mathbf{x}_{13}|) + U(|\mathbf{x}_{23}|).$$

(ii) Each particle contributes separately to the total kinetic energy, thus

$$T = \tfrac{1}{2}m_1|\mathbf{v}_1|^2 + \tfrac{1}{2}m_2|\mathbf{v}_2|^2 + \tfrac{1}{2}m_3|\mathbf{v}_3|^2.$$

(iii) To calculate dU/dt, consider first

$$\frac{dU(|\mathbf{x}_{12}|)}{dt} = \frac{dU(|\mathbf{x}_{12}|)}{d|\mathbf{x}_{12}|} \frac{d|\mathbf{x}_{12}|}{dt}.$$

Now since $|\mathbf{x}_{12}|^2 = \mathbf{x}_{12} \cdot \mathbf{x}_{12}$, differentiating both sides with respect to time, and dividing by 2, we have

$$|\mathbf{x}_{12}|\frac{d|\mathbf{x}_{12}|}{dt} = \mathbf{x}_{12} \cdot (\mathbf{v}_1 - \mathbf{v}_2).$$

Thus

$$\frac{dU(|\mathbf{x}_{12}|)}{dt} = \frac{dU(|\mathbf{x}_{12}|)}{d|\mathbf{x}_{12}|} \frac{\mathbf{x}_{12}}{|\mathbf{x}_{12}|} \cdot (\mathbf{v}_1 - \mathbf{v}_2)$$
$$= -\mathbf{F}_{12} \cdot (\mathbf{v}_1 - \mathbf{v}_2)$$
$$= -\mathbf{F}_{12} \cdot \mathbf{v}_1 - \mathbf{F}_{21} \cdot \mathbf{v}_2.$$

Here use has been made of the relation $\mathbf{F}_{21} = -\mathbf{F}_{12}$. Two similar expressions give the time derivatives of $U(|\mathbf{x}_{13}|)$ and $U(|\mathbf{x}_{23}|)$. Adding all three derivatives together, we have finally

$$\frac{dU}{dt} = -\mathbf{F}_{12} \cdot \mathbf{v}_1 - \mathbf{F}_{21} \cdot \mathbf{v}_2 - \mathbf{F}_{13} \cdot \mathbf{v}_1$$
$$- \mathbf{F}_{31} \cdot \mathbf{v}_3 - \mathbf{F}_{23} \cdot \mathbf{v}_2 - \mathbf{F}_{32} \cdot \mathbf{v}_3$$
$$= -(\mathbf{F}_{12} + \mathbf{F}_{13}) \cdot \mathbf{v}_1 - (\mathbf{F}_{21} + \mathbf{F}_{23}) \cdot \mathbf{v}_2$$
$$- (\mathbf{F}_{31} + \mathbf{F}_{32}) \cdot \mathbf{v}_3$$
$$= -m_1\mathbf{v}_1 \cdot \mathbf{a}_1 - m_2\mathbf{v}_2 \cdot \mathbf{a}_2 - m_3\mathbf{v}_3 \cdot \mathbf{a}_3.$$

In deriving this result, N2 has been assumed to be true for each particle.

For the kinetic energy of particle 1, we have $\tfrac{1}{2}m_1|\mathbf{v}_1|^2 = \tfrac{1}{2}m_1\mathbf{v}_1 \cdot \mathbf{v}_1$, and the time derivative of this is $m_1\mathbf{v}_1 \cdot \mathbf{a}_1$. Adding the contributions from the other two particles, we obtain finally

$$\frac{dT}{dt} = m_1\mathbf{v}_1 \cdot \mathbf{a}_1 + m_2\mathbf{v}_2 \cdot \mathbf{a}_2 + m_3\mathbf{v}_3 \cdot \mathbf{a}_3.$$

It is now clear that

$$\frac{dE}{dt} = \frac{dT}{dt} + \frac{dU}{dt} = 0.$$

This proves that the total energy of the system is conserved.

8.2 Exercises for Units 4–7

Exercise 21 According to Coulomb's law, the force between two point charges is

$$\mathbf{F}_{12} = \frac{q_1 q_2 \mathbf{x}_{12}}{4\pi\varepsilon_0 |\mathbf{x}_{12}|^3},$$

where $\mathbf{x}_{12} = \mathbf{x}_1 - \mathbf{x}_2$. Here we have

$$\mathbf{x}_{12} = (0.4, -0.6, 1)\,\mathrm{m} - (-0.2, 0.2, -1.4)\,\mathrm{m}$$
$$= (0.6, -0.8, 2.4)\,\mathrm{m},$$

and $|\mathbf{x}_{12}| = \sqrt{0.6^2 + 0.8^2 + 2.4^2}\,\mathrm{m} = 2.6\,\mathrm{m}$.
Coulomb's law implies that $\mathbf{F}_{12}$ and $\mathbf{x}_{12}$ are parallel

vectors, and this is clearly true here since $\mathbf{F}_{12} = -15\mathbf{x}_{12}$. It follows from this that

$$\frac{q_1 q_2}{4\pi\varepsilon_0 |\mathbf{x}_{12}|^3} = -15\,\text{N}\,\text{m}^{-1},$$

so that

$$q_2 = \frac{-15 \times (2.6)^3}{9 \times 2.5 \times 10^5}\,\text{C} = -1.17 \times 10^{-4}\text{C}.$$

[Remember that the values of relevant constants are given on the back cover of each Block of S357.]

Exercise 22   The distances of the charges (a)–(e) from the origin are $1.04\,\text{m}$, $0.87\,\text{m}$, $0.99\,\text{m}$, 0 and $1.005\,\text{m}$ respectively. The first and last are clearly outside the sphere, and therefore do not contribute to the total electric flux through its surface. The total of the other three charges is 3.5×10^{-4} C, and, according to Equation 5 Unit 4, the average radial field on the surface of the sphere is thus

$$\frac{3.5 \times 10^{-4}\,\text{C}}{\pi\varepsilon_0} = 3.5 \times 9 \times 10^5\,\text{N}\,\text{C}^{-1}$$

$$= 3.15 \times 10^6\,\text{N}\,\text{C}^{-1}.$$

Exercise 23   (i) The perfect cylindrical symmetry of the system implies that if such a component of the field existed, its magnitude would be constant around the circle. It would therefore be like the field illustrated in Figure 8(a) Unit 4. Such a field is impossible because it could be used to create energy.

(ii) In applying Equation 4, we denote the position of the point P, at which the field is to be determined, as $\mathbf{r}$ instead of $\mathbf{x}$. Thus we have $Q_1 = Q_2 = \lambda\,\Delta x$, $\mathbf{r} = (0, x^2, x^3)$, $\mathbf{x}_1 = -\mathbf{x}_2 = (x, 0, 0)$,

$$|\mathbf{r} - \mathbf{x}_1| = |\mathbf{r} - \mathbf{x}_2| = \sqrt{(x^2)^2 + (x^3)^2 + x^2}$$

$$= \sqrt{r^2 + x^2}.$$

The contribution to the field at P is therefore

$$\frac{\lambda\,\Delta x}{4\pi\varepsilon_0(r^2 + x^2)^{3/2}}\left[(-x, x^2, x^3) + (x, x^2, x^3)\right].$$

This gives the required result immediately. The components parallel to the wire from the two small elements cancel each other.

For completeness, we include the integral of this result over the whole rod, although it was not part of the question. The integral which must be evaluated is

$$\mathbf{E} = \frac{\lambda\mathbf{r}}{2\pi\varepsilon_0}\int_0^\infty \frac{\mathrm{d}x}{(r^2 + x^2)^{3/2}}.$$

This can be simplified by the substitution $x = r\tan\theta$, giving

$$\mathbf{E} = \frac{\lambda\mathbf{r}}{2\pi\varepsilon_0 r^2}\int_0^{\pi/2} \cos\theta\,\mathrm{d}\theta = \frac{\lambda\mathbf{r}}{2\pi\varepsilon_0 r^2}.$$

(iii) Let the length of the cylinder be l. In calculating the total electric flux, Φ, over the cylinder, we see that the ends give no contribution, because there is no component of the field at right-angles to them. The field strength at right-angles to the curved surface of the cylinder is everywhere the same, equal to E, and the area of this surface is $2\pi rl$. The total

charge enclosed by the cylinder is λl, so that

$$\Phi = 2\pi rlE = \lambda l/\varepsilon_0,$$

and this leads to

$$E = |\mathbf{E}| = \frac{\lambda}{2\pi\varepsilon_0 r},$$

as required.

Exercise 24   The direction of the field in each case can be worked out using a sketch. When the field is electric, the acceleration of the test charge is in the direction of the field. In the case of a magnetic field, the acceleration is in the direction of the vector $\mathbf{v} \times \mathbf{B}$. To evaluate this you will need the vector product formula given on the back of each Block.

(i) $\mathbf{E} \propto -\left[(1, 0, 0) - (0, 1, 0)\right] = (-1, 1, 0)$.

$\quad\mathbf{a} \propto \mathbf{E} \propto (-1, 1, 0)$.

(ii) $\mathbf{B} \propto (0, 0, -1)$.

$\quad\mathbf{a} \propto \mathbf{v} \times \mathbf{B} = (1, -1, 2) \times (0, 0, -1) = (1, 1, 0)$.

(iii) $\mathbf{E} \propto (1, 0, 0)$.

$\quad\mathbf{a} \propto \mathbf{E} \propto (1, 0, 0)$.

(iv) Remember that the current is opposite to the electron flow. Use the right-hand rule: place your thumb in the direction of the current, and your fingers curl in the direction of the field.

$\quad\mathbf{B} \propto (0, 1, 0)$.

$\quad\mathbf{a} \propto (1, -1, 2) \times (0, 1, 0) = (-2, 0, 1)$.

(v) Remember that the current is anti-clockwise.

$\quad\mathbf{B} \propto (1, 0, 0)$.

$\quad\mathbf{a} \propto (1, -1, 2) \times (1, 0, 0) = (0, 2, 1)$.

Exercise 25   (i) The particle is moving with non-relativistic velocity, and is acted on by the Lorentz force (Section 3.4 Unit 4) so its acceleration is given by

$$\mathbf{a} = \frac{q}{m}(\mathbf{E} + \mathbf{v} \times \mathbf{B}),$$

$$= (qE^1/m, qv^3 B^1/m, -qv^2 B^1/m)$$

$$= (qE/m, qv^3 B/m, -qv^2 B/m).$$

Because the magnetic part of the Lorentz force is a vector product, the component of $\mathbf{v}$ parallel to the field has no effect on the result.

(ii) At $\mathbf{x} = 0$, $v^2 = 0$ and $v^3 = v$, so that

$$\mathbf{a} = (qE/m, qvB/m, 0).$$

The time taken for the particle to travel a distance d is $t = d/v$, so that applying Equations 14 Unit 1, for uniformly accelerated motion, we have

$$v^2 \approx a^2 t \approx qBd/m,$$

$$x^1 \approx \tfrac{1}{2}a^1 t^2 \approx \frac{qEd^2}{2mv^2}$$

and

$$x^2 \approx \tfrac{1}{2}a^2 t^2 \approx \frac{qBd^2}{2mv}.$$

(iii) The exact expression for **a** obtained in part (i) differs from the approximate value used in part (ii) by an amount which is proportional to v^2. Neglect of this will be a good approximation if $v^2 \ll v$ or

$d \ll mv/qB.$

(iv) Eliminating v between the equations for x^1 and x^2 leads to

$x^2 = Bd\sqrt{qx^1/2mE},$

which is the equation of a parabola. Clearly, for the same B, E and d, the curvature of the parabola depends on the ratio q/m of the particle. Thus particles with different values of this ratio will lie on different curves.

Exercise 26   (i) From Figure 20 Unit 4, it is clear that the magnetic field at **x** is $\mathbf{B} = (0, B, 0)$, where $B = \mu_0 I/2\pi r$. This does not depend at all on x, so irrespective of how far along the wire the test charge has moved, the field acting on it is **B**. The force on the test charge due to the magnetic field is (see Section 3.4 Unit 4) $q\mathbf{v} \times \mathbf{B} = (0, 0, f)$, where

$f = qv^1 B^2 = qvB = \dfrac{\mu_0 q I v}{2\pi r}.$

(ii) It is clear from part (i) that, in the O frame, the test charge is repelled by the wire. There must also be a repulsive force in the O′ frame, as otherwise, the two would observe different phenomena. The magnetic field, **B**′ which is present in the O′ frame cannot be the cause of this force, as the test charge has zero velocity in this frame. It follows that there must be an electric field, **E**′, in the O′ frame and this leads us to expect that the wire is electrically charged.

(iii) The electric field at $(0, 0, r)$ due to an infinite line charge is given by $\mathbf{E}′ = (0, 0, E′)$, where $E′ = \lambda/2\pi\varepsilon_0 r$. The force on the test charge due to this field is $\mathbf{f}′ = q\mathbf{E}′ = (0, 0, qE′)$. This force is in the same direction as **f**, and the ratio of their magnitudes is

$|\mathbf{f}′|/|\mathbf{f}| = E′/vB = \lambda/Iv(\mu_0\varepsilon_0) = \lambda c^2/Iv.$

Exercise 32 concerns the approximate evaluation of λ, and the proof that $\mathbf{f}′ \approx \mathbf{f}$.

Exercise 27   (i) Figure 17 shows the path of the light signal in each of the two inertial frames. In the O frame, the signal travels along the hypotenuses of two right-angled triangles of base $vT/2$ and height d. The total distance travelled is given by Pythagoras' theorem as

$2\sqrt{d^2 + (vT/2)^2} = \sqrt{4d^2 + v^2T^2}.$

In the O′ frame, the signal clearly travels a distance $2d$.

(ii) According to SR2, the velocity of light is c in both frames. It follows that

$\sqrt{4d^2 + v^2T^2}/T = 2d/T′ = c,$

and elimination of d gives immediately

$c^2(T′)^2 + v^2T^2 = c^2T^2.$

From this we find

$T′/T = \sqrt{1 - v^2/c^2} = 1/\gamma,$

which is the required relation.

(iii) T and $T′$ are necessarily equal in the Newtonian world-view.

Exercise 28   (i) The two observers agree about the magnitude of their relative velocity $|v|$. It follows that $|v| = L/T = L′/T′$, and the O′ calculation of the length of the rod is

$L′ = |v|T′ = LT′/T.$

(ii) This problem concerns two events: the arrival of O′ at A and his subsequent arrival at B. These two events both occur at the same place in the O′ frame, i.e. $x′ = 0$, so that the time dilation formula gives immediately $T = \gamma T′$. Inserting this into the result of part (i) gives $L′ = L/\gamma$.

It should be noted that this operational definition of the length of an object as determined by an observer moving relative to it is not quite the same as that used in Section 3.1 Unit 6. There, what the observer does is to locate the ends of the object in his reference frame at the same moment. He then measures the distance between these locations. The Lorentz transformation shows that this always gives the same result as the method used here, where the observer records his own time of transit across the

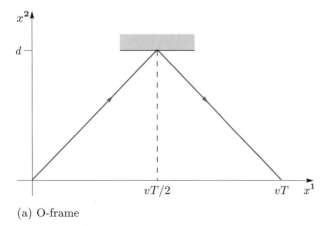

(a) O-frame

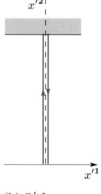

(b) O′-frame

Figure 17

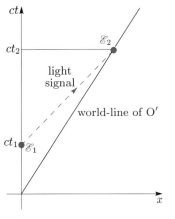
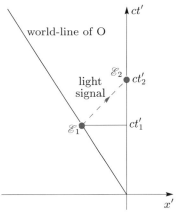

(a) O diagram   (b) O′ diagram

Figure 18

object, then multiplies this by his velocity relative to it. This equivalence is proved in Exercise 35.

Exercise 29   Let the encounters of the two light pulses with the moving object be $\mathscr{E}_a$ and $\mathscr{E}_b$. Referring to Section 6 Unit 5, especially SAQ 25 and the equation above it, we have, for the coordinates of $\mathscr{E}_a$

$$(ct_a, x_a) = (c(8+5)/2, c(8-5)/2) = (6.5c, 1.5c),$$

and for those of $\mathscr{E}_b$

$$(ct_b, x_b) = (c(12+7)/2, c(12-7)/2) = (9.5c, 2.5c).$$

The world-line of a particle moving with uniform velocity between these events is

$$
\begin{aligned}
x &= \left(\frac{t-t_a}{t_b-t_a}\right)(x_b - x_a) + x_a \\
&= \left(\frac{t-6.5}{9.5-6.5}\right)(2.5c - 1.5c) + 1.5c \\
&= \frac{c}{3}(t-2).
\end{aligned}
$$

Exercise 30   (i) The sketch drawn by O is shown in Figure 18(a) and that of O′ in Figure 18(b). In 18(a), the slope of the O′ world-line is c/v, and in 18(b) the slope of the O world-line is $-c/v$. The slope of the world-line for the light signal in both diagrams is 1.

(ii) Referring to Figure 18(a), the equation of the O′ world-line is $x = vt$, and that of the light signal is $x = c(t - t_1)$. (It has a slope of 1 and passes through the point $(0, ct_1)$.) These must intersect at $t = t_2$, so that $vt_2 = c(t_2 - t_1)$, and this leads to

$$t_2 = t_1/(1 - v/c).$$

Similarly, in Figure 18(b), the world-line of O is $x' = -vt'$, while for the light signal $x' = c(t' - t_2')$. (Again, it has a slope of 1 and passes through $(0, ct_2')$.) By the same argument as before, these intersect at $t' = t_1'$. Thus $-vt_1' = c(t_1' - t_2')$, hence

$$t_1' = t_2'/(1 + v/c).$$

(iii) For O′, $\mathscr{E}_0$ and $\mathscr{E}_2$ occur at the same place, i.e. $x' = 0$, and time dilation therefore gives $t_2/t_2' = \gamma$.

The ratio t_1/t_1' is given by

$$t_1/t_1' = (1 - v^2/c^2)t_2/t_2' = 1/\gamma.$$

This *is* consistent with the time dilation formula because $\mathscr{E}_0$ and $\mathscr{E}_1$ occur at the same place for O.

Exercise 31   (i) For $v = \frac{45}{53}c$, the time-dilation factor is $\gamma = \frac{53}{28}$, so that the Lorentz transformation for this case is

$$x' = \tfrac{53}{28}x - \tfrac{45}{28}ct \qquad ct' = \tfrac{53}{28}ct - \tfrac{45}{28}x.$$

$\mathscr{E}_1$ is the event $(0, 56)\,\mathrm{m}$ for O, while for O′ it is

$$(-45 \times 56/28, 53 \times 56/28)\,\mathrm{m} = (-90, 106)\,\mathrm{m}.$$

Thus $\mathscr{E}_1$ occurs at $ct' = ct_1' = -90\,\mathrm{m}$, and the length of the train, (i.e. the separation of $\mathscr{E}_0$ and $\mathscr{E}_1$ in the rest frame of the train) is $106\,\mathrm{m}$. The length of the train can also be obtained directly from the length contraction formula.

(ii) The spacetime diagram drawn from the viewpoint of O is shown in Figure 19(a). The world-lines of the ends of the tunnel are vertical lines, separated by $56\,\mathrm{m}$ on the x-axis. The world-lines of the ends of the train are parallel lines of slope $\frac{53}{45}$. The events $\mathscr{E}_0$ and $\mathscr{E}_1$ occur at the intersections of the two pairs of world-lines as shown. According to O, both events occur at $t = 0$, the tunnel and train are the same length (i.e. $56\,\mathrm{m}$), and the world-line of the engine is

$$x = \tfrac{45}{53}ct + 56.$$

The spacetime diagram from the viewpoint of O′ is shown in Figure 19(b). The world-lines of the rear coach and the engine are now vertical and cross the space axis at $x' = 0$ and $x' = 106\,\mathrm{m}$ respectively. The world-lines of the mouth and end of the tunnel are now parallel lines of slope $-\frac{53}{45}$. As before, the events $\mathscr{E}_0$ and $\mathscr{E}_1$ occur at the intersections of the two pairs of world-lines, but now $\mathscr{E}_1$ is $(-90, 106)\,\mathrm{m}$. The world-line of the tunnel end is thus the straight line passing through $(-90, 106)$ and having slope $-\frac{53}{45}$, which is

$$\frac{ct' + 90}{x' - 106} = -\frac{53}{45}.$$

After some rearrangement, this becomes

$$x' = -\tfrac{45}{53}ct' + \tfrac{1568}{53}.$$

(a) O

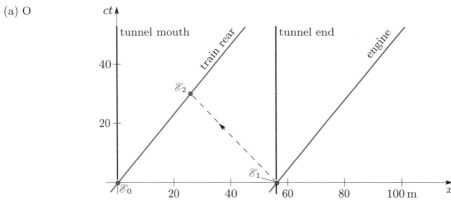

(b) O′

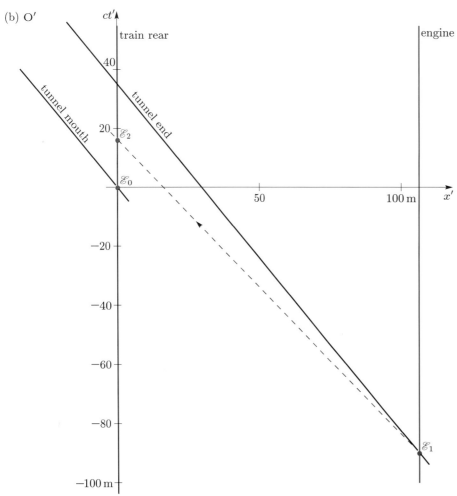

(c) O-viewpoint

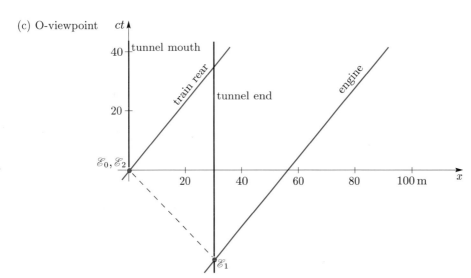

Figure 19

The length of the tunnel as observed by O' is given by the intersection of this line with the space axis $ct' = 0$, and is therefore $1568/53 = 29.58$ m. (*Note*: In such questions you can always use the Lorentz transformation directly, so the length of the tunnel according to an observer on the train is just length$/\gamma = 56/1.893 = 29.58$ m. See the boxed tutorial at the beginning of Section 5.3.)

These results may be summarized by saying that, whereas for O, $\mathscr{E}_0$ and $\mathscr{E}_1$ are simultaneous, and the train and the tunnel are the same length, for O', $\mathscr{E}_1$ occurs before $\mathscr{E}_0$ and the train is much longer than the tunnel.

(iii) O' will see the engine emerge from the end of the tunnel when a light signal reaches him from event $\mathscr{E}_1$. This light signal is shown in both Figures 19(a) and 19(b), and $\mathscr{E}_2$ is the intersection of its world-line with that of O'. Using the O reference frame, the O' world-line is $x = \frac{45}{53}ct$ and that of the light signal is $x = 56 - ct$. These lines intersect when $ct = ct_2 = 212/7$ m and $x = x_2 = 180/7 = 25.71$ m. Thus event $\mathscr{E}_2$ is $(212/7, 180/7)$ m in the O frame, and O' is 25.71 m into the tunnel when it occurs.

In the O' frame, the world-line of the light signal is the straight line of slope -1 passing through $(-90, 106)$ m, i.e.

$$\frac{ct' + 90}{x' - 106} = -1,$$

and this simplifies to $x' = -ct + 16$. The intersection of this with the O' world-line, $x' = 0$, gives $ct' = ct_2' = 16$ m for the time of $\mathscr{E}_2$ in this frame.

(iv) The motion of the train through the second tunnel is shown in Figure 19(c), which is drawn from the O viewpoint. (*Note*: O is now an observer stationed at the mouth of the second tunnel.) Now $\mathscr{E}_2$ and $\mathscr{E}_0$ are the same event, $(0, 0)$. The world-line of the light pulse is $x = -ct$, and $\mathscr{E}_1$ occurs at the intersection of this with the world-line of the engine, given above, so that

$$x_1 = -ct_1 = \frac{45}{53}ct_1 + 56.$$

Thus $\mathscr{E}_1$ is the event $(ct_1, x_1) = (-212/7, 212/7)$ m, and the length of the tunnel is $212/7 = 30.29$ m.

Exercise 32   (i) We can picture the wire as consisting of two superimposed infinite line charges, one with a positive charge density $\lambda^+ = ne$, and one with negative charge density, λ^-, which moves with velocity u along the positive **1**-axis. Consider a length l of the latter. It contains a total charge $l\lambda^-$, and in the O frame, this same quantity of charge is observed to be contained in a length $l\sqrt{1 - u^2/c^2}$. The observed charge density is thus $\lambda^-/\sqrt{1 - u^2/c^2}$, and this must cancel the effect of the positive line charge. Thus

$$\lambda^- = -ne\sqrt{1 - u^2/c^2}.$$

(ii) In the O frame, the positive line charge moves with velocity v, and the negative line charge moves with velocity u'. By the same argument as in part

(i), the total charge density in this frame is

$$\lambda = \frac{ne}{\sqrt{1 - v^2/c^2}} + \frac{\lambda^-}{\sqrt{1 - u'^2/c^2}}$$

$$= ne\left[\frac{1}{\sqrt{1 - v^2/c^2}} - \frac{\sqrt{1 - u^2/c^2}}{\sqrt{1 - u'^2/c^2}}\right].$$

This result shows that, as anticipated in SAQ 6 Unit 5, the charge imbalance arises from two separate effects: an increase in the density of positive charge, together with a decrease in the density of negative charge. Putting $u' = 0$ and $u = v$, which is the case considered in Unit 5, the two effects are easily seen to be equal up to order v^2/c^2.

(iii) Making the indicated approximations, we have

$$\lambda \approx ne[1 + \tfrac{1}{2}v^2/c^2 - (1 - \tfrac{1}{2}u^2/c^2)(1 + \tfrac{1}{2}(u - v)^2/c^2)$$

$$\approx \tfrac{1}{2}ne[v^2/c^2 + u^2/c^2 - (u - v)^2/c^2] = neuv/c^2.$$

(iv) Since $I = neu$, the line charge density is $\lambda \approx Iv/c^2$. From the last part of Exercise 26, and using this last result,

$$|\mathbf{f}'|/|\mathbf{f}| = \lambda c^2/Iv \approx 1.$$

It follows that $\mathbf{f}' \approx \mathbf{f}$. Note that, from the discussion in Section 3.5 Unit 4, these two forces are not expected to be exactly equal. The exact result, i.e. without the approximations made in part (iii), turns out to be $\mathbf{f}' = \mathbf{f}/\sqrt{1 - v^2/c^2}$.

Exercise 33   (i) Using the inverted form of the Lorentz transformation

$$x = \gamma(x' + vt') \qquad t = \gamma(t' + vx'/c^2),$$

and substituting for x and t, we have

$$x/\lambda - ft = \gamma(x' + vt')/\lambda - \gamma f(t' + vx'/c^2).$$

Making use of $\lambda f = c$, the r.h.s. becomes

$$\gamma(1 - v/c)x'/\lambda - \gamma f(1 - v/c)t',$$

and this is of the form $x'/\lambda' - f't'$, if we identify λ' and f' with the following expressions

$$\lambda' = \lambda/\gamma(1 - v/c) = \lambda\sqrt{\frac{1 + v/c}{1 - v/c}},$$

and

$$f' = f\gamma(1 - v/c) = f\sqrt{\frac{1 - v/c}{1 + v/c}}.$$

These results are in agreement with the Doppler shift formula (Equation 5 Unit 6).

(ii) It is easily seen that $\lambda'f' = \lambda f = c$. The product of frequency and wavelength is the wave velocity, and by SR2 this will be the same for all observers.

Exercise 34   (i) The world-line of O' is $x = 5ct/13$ and that of the light signal is $x = ct - 3$. Event $\mathscr{E}_2$ occurs at the intersection of these two lines, which is at $(ct_2, x_2) = (4.875, 1.875)$. For $v/c = 5/13$, $\gamma = 13/12$, so that

$$ct_2' = ct_2 \times 12/13 = 4.875 \times 12/13 = 4.5.$$

This result follows from the time-dilation formula (Equation 26 Unit 6) as both events $\mathscr{E}_0$ (the coincidence of the origins of O and O') and $\mathscr{E}_2$ occur at the origin of O'. As a check on this, we note that the Lorentz transformation between the O and O' frames is

$$x' = \tfrac{13}{12}x - \tfrac{5}{12}ct \qquad ct' = \tfrac{13}{12}ct - \tfrac{5}{12}x,$$

and substituting in the values of (ct_2, x_2) found earlier, gives immediately $(ct_2', x_2') = (4.5, 0)$.

(ii) From Equation 29, the velocity of the rocket relative to O is

$$\frac{5c/13 - 4c/5}{1 - 4/13} = -\frac{3}{5}c.$$

The distance which the rocket travels to reach O is 1.875, so the time taken, according to the O clock, is $1.875 \times 5/3c = 3.125/c$. Thus

$$t_3 = t_2 + 3.125/c = 4.875/c + 3.125/c = 8/c.$$

The sequence of events is shown from the O viewpoint in Figure 20.

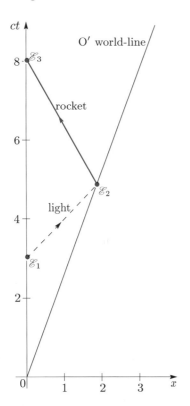

Figure 20

(iii) For the rocket's motion relative to O, $\gamma = 1/\sqrt{1 - 3^2/5^2} = 5/4$. Thus the time-dilation formula gives the time elapsed between $\mathscr{E}_2$ and $\mathscr{E}_3$ on the clock aboard the rocket as $3.125 \times 4/5c = 2.5/c$. For the first part of the journey, from $\mathscr{E}_0$ to $\mathscr{E}_2$, the rocket's clock agrees with that of O'. It follows that the total time between $\mathscr{E}_0$ and $\mathscr{E}_3$ recorded by the clock aboard the rocket is $4.5/c + 2.5/c = 7/c$. This is clearly *less* than the total time between these events recorded by O, which is $8/c$. The result is in accordance with Section 4.2, if we regard O as the twin who stays at home, while his twin brother travels with the rocket.

Exercise 35 The spacetime diagram is shown in Figure 21. The world-lines of the two ends, A and B, of the rod are vertical and pass through $x = 0$ and $x = L$. The events C and D are the intersections of the world-line of B with the x'- and ct'-axes, respectively, of O'. At $t' = 0$, the rod occupies the segment AC, and this clearly represents its length, L', in the O' frame. The ct'-axis is the world-line of O', so that D is the event at which O' reaches end B of the rod. It follows that the segment AD represents cT', where T' is the O' transit time along the rod. A glance at Figure 9 of Unit 7 indicates that the scales on the x'- and ct'- axes are equal. (The figure is symmetric about the world-line of the light pulse, and the two sets of parallel lines have the same spacing.) Thus

$$\frac{L'}{cT'} = \frac{\text{AC}}{\text{AD}}.$$

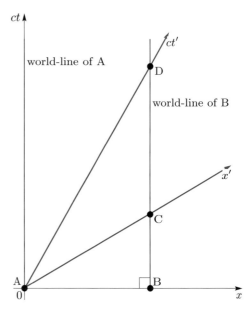

Figure 21

As stated in Section 4.1 of Unit 7, the lines of constant x' (including the ct'-axis) all have slope c/v, while those of constant ct' (including the x'-axis) have slope v/c. This implies that

$$\frac{\text{BC}}{\text{AB}} = \frac{\text{AB}}{\text{BD}} = \frac{v}{c},$$

and therefore, by Pythagoras' theorem,

$$\text{AC} = \sqrt{\text{BC}^2 + \text{AB}^2} = \frac{v}{c}\sqrt{\text{AB}^2 + \text{BD}^2} = \frac{v}{c}\text{AD}.$$

It now follows that

$$\frac{\text{AC}}{\text{AD}} = \frac{v}{c},$$

and hence that $L' = vT'$, which is the required result.

Exercise 36 (i) The γ factor for $v/c = 3/5$ is $5/4$. Thus the Lorentz transformation is

$$x' = \tfrac{5}{4}x - \tfrac{3}{4}ct \qquad ct' = \tfrac{5}{4}ct - \tfrac{3}{4}x.$$

(ii) The world-line of particle P is $x = \tfrac{4}{5}ct$. The release of particle Q is the event $(ct', x') = (0, 4)$ in the O′ frame. Using the inverted Lorentz transformation, we find

$$x = \tfrac{5}{4}x' + \tfrac{3}{4}ct' = 5, \qquad ct = \tfrac{5}{4}ct' + \tfrac{3}{4}x' = 3.$$

Thus the release of Q is $(ct, x) = (3, 5)$ in the O frame. The world-line of Q is therefore

$$x - 5 = \frac{-1}{2}(ct - 3), \quad \text{or} \quad x = -\tfrac{1}{2}ct + \tfrac{13}{2}.$$

P and Q collide when

$$\tfrac{4}{5}ct = -\tfrac{1}{2}ct + \tfrac{13}{2},$$

i.e. at $ct = 5$. The position of the collision is given by $x = \tfrac{4}{5} \times 5 = 4$. Thus in the O frame, the collision is the event $(5, 4)$. Substituting these coordinates into the Lorentz transformation gives

$$x' = \tfrac{5}{4} \times 4 - \tfrac{3}{4} \times 5 = \tfrac{5}{4}$$

and

$$ct' = \tfrac{5}{4} \times 5 - \tfrac{3}{4} \times 4 = \tfrac{13}{4},$$

so that for O′, the collision is the event $(3.25, 1.25)$.

Exercise 37
(i) From Theorem I, we have

$$15^2 c^2 - 0^2 = 17^2 c^2 - (x'_b - x'_a)^2,$$

so that

$$|x'_b - x'_a| = \sqrt{17^2 - 15^2}\,c = 8c.$$

Only the magnitude of the space separation can be determined, not the sign. Clearly, according to O′

$$(c(t'_b - t'_a), x'_b - x'_a) = (17c, \pm 8c).$$

(ii) There is no loss of generality if O and O′ reset their clocks to zero at $\mathscr{E}_a$. Thus in the O′ frame, $\mathscr{E}_b$ is $(17c, \pm 8c)$. He reverses his direction of motion at the event $(7c, 0)$ (call this $\mathscr{E}_c$), so that he moves along two straight world-lines, from $\mathscr{E}_a$ to $\mathscr{E}_c$, then from $\mathscr{E}_c$ to $\mathscr{E}_b$. The total proper time elapsed is, according to the definition in Section 5 Unit 7,

$$\sqrt{(t'_c - t'_a)^2 - \frac{(x'_c - x'_a)^2}{c^2}} + \sqrt{(t'_b - t'_c)^2 - \frac{(x'_b - x'_c)^2}{c^2}}$$

$$= 7 + \sqrt{(17 - 7)^2 - 8^2}\ \text{s} = 13\,\text{s}.$$

This is certainly less than the 15 s recorded by O, as required by Theorem II.

(iii) O′ can keep his appointment with O by adjusting his return velocity, as long as his separation from $\mathscr{E}_b$ is *time-like*, i.e. it is in his absolute future (Region I, Figure 11 Unit 7). Thus suppose, more generally, $\mathscr{E}_c$ were the event $(ct', 0)$ in the O′ frame, then the relevant condition is

$$S_{cb} = (17 - t')^2 c^2 - 8^2 c^2 \geqslant 0,$$

from which it follows that $t' \leqslant 9$. Thus the limiting time is 9 s, after which he would have to complete his journey at velocity c. If he delays any more, he will not reach $\mathscr{E}_b$ at all.

Exercise 38
(i) The Lorentz transformation, inverted to give the O spacetime coordinates in terms of those of O′, is

$$x = \gamma(x' + vx'^0/c) \qquad x^0 = \gamma(x'^0 + vx'/c)$$

(cf. Equations 30a, 30b Unit 7). Replacing (x^0, x) by (p^0, p), we find

$$p = \gamma(p' + vp'^0/c) \qquad p^0 = \gamma(p'^0 + vp'/c).$$

(ii) A particle at rest in the O′ frame, and therefore moving at velocity v relative to O, has zero linear momentum according to O′, i.e. $p' = 0$. In this case, the equations obtained in part (i) give immediately

$$p = \gamma p'^0 v/c \qquad p^0 = \gamma p'^0.$$

(iii) If v is very small compared to c, then $\gamma \approx 1$ and $p \approx p'^0 v/c$. In this non-relativistic limit, p should agree with the Newtonian formula, $p = mv$, where m is the mass of the particle. This is seen to be the case if $p'^0 = mc$. Substituting this result into the formulae obtained in part (ii), we have finally

$$p = \gamma mv \qquad p^0 = \gamma mc.$$

(iv) In one dimension, $\mathbf{p} = (p, 0, 0)$, $\mathbf{v} = (v, 0, 0)$, and

$$1/\sqrt{1 - |\mathbf{v}|^2/c^2} = 1/\sqrt{1 - v^2/c^2} = \gamma.$$

The results obtained in part (iii) are therefore consistent with the assignments for $\mathbf{p}$ and p^0 made by O and O′.

Exercise 39
(i) The pion has energy E_π and the proton is at rest, so that $P^0 = E_\pi/c + m_p c$. The linear momentum of the proton is clearly zero, and hence $\mathbf{P} = \mathbf{p}_\pi$. An expression for E_π is given immediately by Equation 36 Unit 7, namely $E_\pi^2/c^2 - |\mathbf{p}_\pi|^2 = m_\pi^2 c^2$, from which we obtain

$$E_\pi = \sqrt{m_\pi^2 c^4 + |\mathbf{p}_\pi|^2 c^2}.$$

(ii) In any other inertial frame, the components of the four-vector $(P^0, \mathbf{P})$ would be different, say $(P'^0, \mathbf{P}')$, and the two sets of components would be related by a Lorentz transformation. It follows that

$$(P^0)^2 - |\mathbf{P}|^2 = (P'^0)^2 - |\mathbf{P}'|^2 = M^2 c^2,$$

where M is a constant. This is what is meant by the statement that this quantity is a Lorentz invariant. The result is stated, for a single particle, in Theorem III (Unit 7), and we have already used it for the pion in part (i). However, it applies equally well to a system of two particles. From the result of part (i), we have

$$\begin{aligned}(P^0)^2 - |\mathbf{P}|^2 &= (E_\pi/c + m_p c)^2 - |\mathbf{p}_\pi|^2 \\ &= E_\pi^2/c^2 + 2E_\pi m_p + m_p^2 c^2 - |\mathbf{p}_\pi|^2 \\ &= m_\pi^2 c^2 + 2E_\pi m_p + m_p^2 c^2.\end{aligned}$$

(iii) If E is the energy of the system in the centre-of-momentum frame, then, since the total momentum in this frame is zero, the value of the above Lorentz invariant is E^2/c^2, and this may be set equal to the result in part (ii), giving immediately

$$E = \sqrt{m_\pi^2 c^4 + 2E_\pi m_p c^2 + m_p^2 c^4}.$$

(iv) As the colliding particles have zero total momentum, the two strange particles produced in the collision must also have no net momentum. This implies that they must depart in opposite directions with their momenta exactly cancelling. It is therefore clear that the minimum energy is achieved by *producing them both at rest*. The total energy of these particles in this situation is $E = m_\Lambda c^2 + m_K c^2$, and this must be equal to the total energy of the particles which originally collided. The threshold energy for the reaction is therefore

$$E_\pi = \frac{(m_\Lambda c^2 + m_K c^2)^2 - m_\pi^2 c^4 - m_p^2 c^4}{2 m_p c^2}$$

$$= \frac{(1116 + 498)^2 - 140^2 - 938^2}{2 \times 938} \text{ MeV}$$

$$= 909 \text{ MeV}.$$

If the incident pion has less energy than this, the Λ^0 and K^0 are not produced. The factor γ for the pion is given by

$$\gamma = \frac{E_\pi}{m_\pi c^2} = \frac{909}{140} = 6.493.$$

Its speed is therefore

$$v = c\sqrt{1 - 1/\gamma^2} = 0.988c.$$

Exercise 40   (i) The momentum and energy of a particle of mass m are given in terms of the velocity, $\mathbf{v}$, by

$$\mathbf{p} = m\gamma\mathbf{v} \quad \text{and} \quad E = p^0 c = m\gamma c^2,$$

where $\gamma = 1/\sqrt{1 - v^2/c^2}$ and $v = |\mathbf{v}|$ (cf. Exercise 38). These formulae can be rearranged easily to give the velocity and energy in terms of the momentum. From the momentum equation, we have

$$|\mathbf{p}|^2 = (m\gamma v)^2 = m^2 c^2 (\gamma^2 - 1),$$

and turning this round gives $mc\gamma = \sqrt{m^2 c^2 + |\mathbf{p}|^2}$, from which the velocity is given by

$$\mathbf{v} = \frac{\mathbf{p}}{m\gamma} = \frac{\mathbf{p}c}{\sqrt{m^2 c^2 + |\mathbf{p}|^2}}.$$

Similarly, for the energy we have

$$E = \gamma m c^2 = c\sqrt{m^2 c^2 + |\mathbf{p}|^2}.$$

(ii) If $m = 0$, we have $\mathbf{v} = c\mathbf{p}/|\mathbf{p}|$ and $E = |\mathbf{p}|c$, or $p^0 = |\mathbf{p}|$. Thus a particle of zero mass travels at the velocity of light, and its energy is equal to the magnitude of its momentum, multiplied by c.

(iii) Eliminating the square root factor between the two equations derived in part (i) leads immediately to $\mathbf{v} = \mathbf{p}c^2/E$. This relation is often useful in the form

$$\mathbf{p} = \frac{E}{c^2}\mathbf{v}.$$

Exercise 41   (i) As the particle has zero mass, its momentum components in the O frame are

$$p^0 = |\mathbf{p}| = \frac{E}{c}$$

and

$$\mathbf{p} = (p^1, p^2, p^3) = \left(\frac{E}{c}\cos\theta, \frac{E}{c}\sin\theta, 0\right).$$

(ii) In the O′ frame the momentum components are

$$p'^0 = |\mathbf{p}'| = \frac{E'}{c}$$

and

$$\mathbf{p}' = (p'^1, p'^2, p'^3) = \left(\frac{E'}{c}\cos\theta', \frac{E'}{c}\sin\theta', 0\right),$$

the two sets of components being related by the Lorentz transformation as given in Equations 23 of Unit 6. In particular, we have for the time components

$$p'^0 = \gamma(p^0 - vp^1/c),$$

which is equivalent to

$$E' = \gamma E\left(1 - \frac{v}{c}\cos\theta\right).$$

(iii) The **1**- and **2**-components of momentum in the two frames are related by $p'^1 = \gamma(p^1 - vp^0/c)$ and $p'^2 = p^2$. Thus

$$\tan\theta' = \frac{p'^2}{p'^1} = \frac{p^2}{\gamma(p^1 - vp^0/c)} = \frac{\sin\theta}{\gamma(\cos\theta - v/c)}.$$

Although it was not part of the question, it is worth noting that it is usually more useful to have the expression for E' in terms of the actual, 'aberrated' angle of observation, θ'. This can be obtained by the familiar trick of interchanging O and O′, and reversing the sign of v. Thus

$$E = E'\gamma\left(1 + \frac{v}{c}\cos\theta'\right),$$

from which the result

$$E' = E/\gamma\left(1 + \frac{v}{c}\cos\theta'\right)$$

follows immediately.

Exercise 42   (i) In the rest-frame of the pion, the total momentum of the system is zero. (It is the centre-of-momentum frame!) This means that the momenta of the photons must be equal and opposite, so that they must each have energy $\frac{1}{2}m_\pi c^2$. These photons obviously travel in opposite directions (or else momentum would not be conserved.)

(ii) By conservation of momentum and energy, the momentum four-vector of the photon is $((E_1 + E_2)/c, \mathbf{p}_1 + \mathbf{p}_2)$, and from Exercise 40 part (ii), we have

$$\frac{c|\mathbf{p}_1 + \mathbf{p}_2|}{E_1 + E_2} = 1.$$

(iii) The *maximum* value of $|\mathbf{p}_1 + \mathbf{p}_2|$ occurs when the two momentum vectors are parallel, and is equal to

$$|\mathbf{p}_1| + |\mathbf{p}_2| = (E_1|\mathbf{v}_1| + E_2|\mathbf{v}_2|)/c^2.$$

In deriving this, the result of Exercise 40 part (iii) has been used. The positron and electron both have non-zero mass and therefore

$$\frac{c|\mathbf{p}_1 + \mathbf{p}_2|}{E_1 + E_2} \leqslant \frac{E_1|\mathbf{v}_1|/c + E_2|\mathbf{v}_2|/c}{E_1 + E_2} < 1,$$

since $|\mathbf{v}_1|/c$ and $|\mathbf{v}_2|/c$ are both < 1. This means that the reaction cannot conserve momentum and energy, and therefore cannot occur. The specific difficulty here is that the photon carries a huge amount of momentum for a given energy, and the two particles cannot carry away so much momentum, even when moving in the same direction. The pair production process does occur in matter, however, when a nearby atomic nucleus, by recoiling, absorbs the excess momentum. This nuclear recoil absorbs very little energy because of the large mass of the nucleus.

9 Example answers to the essay exercises

1 The principles of homogeneity and isotropy of space assert that the laws of physics do not distinguish one position or direction in space from another. They form part of the principle of relativity.

(i) These two principles imply that the location and orientation of the coordinate system may be chosen arbitrarily. A rigid rod serves as a measuring stick, and may be used to calibrate all three axes. By assumption, its length is the same whatever its position or orientation. The orthogonality of each pair of axes is established by means of a Pythagorean triangle using the same rod to check the length of the hypotenuse.

(ii) Ptolemy regarded the centre of the Earth as a special position in space, and the orientations of his epicycles picked out special directions in space. Copernicus thought of the Earth as orbiting the Sun, but his description of planetary orbits was in terms of cycles and epicycles centred at the centre of the Earth's orbit, again implying a privileged position. Kepler's empirical laws accurately summarize the basic facts of planetary motion, and place the Sun in a special position at the focus of each elliptical orbit. Newton's theory explains these empirical facts in terms of the gravitational attraction of each planet by the Sun. This view is consistent with the homogeneity and isotropy of space because it implies that the Solar System, regarded as an approximately isolated system, would behave in an identical manner in any location or orientation in space.

(iii) These conservation laws state that

$$\frac{d\mathbf{P}}{dt} = \mathbf{0} \qquad \frac{d\mathbf{J}}{dt} = \mathbf{0},$$

where, for an isolated two-particle system, the momentum, $\mathbf{P}$, and the angular momentum, $\mathbf{J}$, are defined as

$$\mathbf{P} = m_1\mathbf{v}_1 + m_2\mathbf{v}_2 \qquad \mathbf{J} = m_1\mathbf{x}_1 \times \mathbf{v}_1 + m_2\mathbf{x}_2 \times \mathbf{v}_2.$$

Conservation of momentum follows if the potential energy function, U, depends only on $\mathbf{x}_1 - \mathbf{x}_2$. This ensures that U is value and form invariant under spatial translations, and thus expresses the homogeneity of space. In a similar way, conservation of angular momentum follows if U depends only on $|\mathbf{x}_1 - \mathbf{x}_2|$. This expresses the isotropy of space through the value and form invariance of U under spatial rotations.

2 (i) The first law asserts that inertial frames exist, and extend over all space. An inertial frame is defined as one in which a given isolated free particle moves in a straight line at constant speed. But beyond that, N1 declares that in a frame so defined, *all* free particles will move in straight lines at constant speed.

(ii) N2 states that the acceleration, $\mathbf{a}$, of a body due to a force $\mathbf{F}$, is $\mathbf{F}/m$, where m is its mass. The magnitude of $\mathbf{a}$, is the same for all inertial observers. If the same force acts on two different bodies, N2 implies that the ratio of their accelerations is constant, no matter what the strength or nature of the force. This can be used to define the mass of any body by determining the ratio of its mass to that of a standard body.

If a force, $\mathbf{F}$, induces an acceleration, $\mathbf{a}_1$, in a body of mass m_1, and an acceleration $\mathbf{a}_2$ in a body of mass m_2, then, if we assume that mass is *additive*, the acceleration, $\mathbf{a}$, induced by $\mathbf{F}$ in the composite body (i.e. the two stuck together) satisfies

$$\frac{1}{|\mathbf{a}|} = \frac{1}{|\mathbf{a}_1|} + \frac{1}{|\mathbf{a}_2|}.$$

This synthesis of N2 and additivity of mass is thus capable of experimental verification.

Moreover, if we assume the superposition principle for forces, then N2 implies that two forces acting together on a body will cause an acceleration which is the vector sum of the accelerations each force will cause if acting independently on the same body. This also can be tested experimentally.

(iii) N1 would be disproved if an isolated free particle were observed to accelerate in an inertial frame of reference.

N2 can be tested by applying a variety of forces in turn to two bodies and verifying that their accelerations are in a constant ratio. This ratio gives the inverse ratio of the inertial masses of the bodies. In addition, as indicated in part (ii), when N2 is taken together with other, reasonable assumptions, namely additivity of mass and superposition of forces, it provides further experimentally verifiable (or falsifiable) predictions.

3 The principle of relativity states that two equally calibrated, inertial frames, S and S′, cannot be distinguished from each other by any (mechanical) experiment. This means that if identical sets of particles, isolated from the rest of the universe, are observed in S and S′, and if the initial positions and velocities are the same in both frames, the evolution of both systems will be the same. This means that, at any subsequent time, the positions and velocities of corresponding particles will be the same in S and S′.

An inertial frame is one in which a free particle, i.e. one which is not acted on by any force, has zero acceleration. The principles of homogeneity of space and time, and isotropy of space imply that if S is inertial and S′ differs from S in that its origin is at a different place, its axes orientated in different directions, or its clock was started at a different instant, then S′ is necessarily inertial also. S′ will also be inertial if its origin is moving at constant speed along a straight line in S, and its axes are not rotating relative to those of S.

The three objections all suffer from the same defect: they do not refer to isolated systems. Thus in (i), a system of particles close to the Earth is not isolated unless the Earth itself is included. The same is true for a system of particles close to the Moon. Thus the two systems are not identical, and their behaviour is inevitably different.

Weather at a point on the Earth's surface is affected not by time itself, but by the elevation of the Sun above the horizon at noon. This depends on the solar declination, which varies throughout the year. The statement in (ii) is false because any system excluding the Sun is not isolated. An experiment consistent with the conditions, stated above for the validity of the principle of relativity, could only compare weather at times when the Sun and Earth are in the same relationship to each other.

In (iii), the Earth is excluded from the system, and this falsifies the conclusion. An object falls towards the centre of the Earth whatever the orientation in space of this direction.

4 The new picture referred to is Einstein's special theory of relativity (SR). Observation is the process of assigning positions and times to events in an inertial reference frame. The inertial frame, in principle, extends over all space, and a method must be established of observing events occurring anywhere. This can be done, for example, by employing a set of data-takers, distributed through space and supplied with identical synchronized clocks. Any sequence of events can be studied subsequently when each data-taker reports the times of events occurring close to its own location.

The length of a moving (metre) rod can be measured in the following way. Suppose the leading end of the rod passes a certain data-taker at 12 noon, and its trailing end passes another data-taker at the same moment. The length of the rod is simply the distance between these two data-takers. The length contraction phenomenon in SR tells us that it is necessarily less than 1 metre.

Similarly, suppose a moving clock which ticks at 1 second intervals according to an observer moving with it, passes a given data-taker at 12 noon just as it ticks. It passes a second data-taker exactly on its next tick. Time dilation in SR implies that the time registered on the second data-taker's clock at this moment will be later than 1 second after noon.

The time at which an event is *observed* to occur in an inertial frame is the time registered for the event by a data-taker at the location where it occurs. This is quite different from the time at which the event is *seen* from a position a distance d away. The difference is d/c, the transit time of light across the intervening space. The visual image at any moment is made up of light signals which happen to reach the eye simultaneously. However, the events from which these signals originated, would have been simultaneous only if they had occurred at equal distances from the eye.

5 (i) Postulates A1 to A7 are common to both viewpoints. In special relativity also, space and time are passive, homogeneous and continuous; space is isotropic and complete. What about A8? The answer is a little more subtle. Space and time become 'mixed up' when one moves from one inertial frame to another, and Minkowski replaced them as separate entities by 'spacetime'; thus measured quantities of time and distance are certainly not independent. Nevertheless, because space and time each have their distinctive ways of being measured, and moreover enter into the expression for the invariant interval in distinctly different ways, there is a sense in which they remain 'independent concepts'.

Postulates A9 and A10, though regarded as self-evident in the older view, are not true in special relativity. They fail because, in the new picture, different inertial observers do not, in general, agree about either the distance, Δx, or the time interval, Δt, between two events. However, they do agree about the value of the invariant interval, $c^2(\Delta t)^2 - (\Delta x)^2$.

(ii) The principle of relativity, as formulated by Galileo, is affirmed in special relativity. Galilean relativity (together with the Galilean transformation) is a basic principle of Newtonian mechanics, but fails when applied to electromagnetism, a branch of physics unknown to Galileo and Newton. Einstein's two postulates assert that the principle is true for *all* physical laws, but imply that the spacetime coordinates of an event, as determined by different inertial observers, are related through the Lorentz rather than the Galilean transformation.

(iii) The concept of momentum carries over into special relativity, but it now possesses a 'time' component and forms a 'four vector', $(E/c, \mathbf{p})$, analogous to $(ct, \mathbf{x})$. As in Newtonian mechanics, the total linear momentum, $\mathbf{p}_1 + \mathbf{p}_2$, of a pair of particles is the same after the collision as before. The total energy, $E_1 + E_2$, is also conserved in the collision, but here there is a very important difference from Newtonian mechanics. The energy of a particle does not vanish at zero velocity, but takes the value mc^2. This implies that the particle has energy by virtue of its mass, and that, while the total energy is conserved, the sum of the masses of the particles is not. This leads to the possibility that the particles emerging from a collision may be quite different in nature and number from the original participants.

Appendix: Writing essays for TMAs and the examination

There will be an essay question in the examination and also in at least one TMA. We hope that the TMA essay will prepare you for the exam essay. Naturally, writing an essay for an exam is not quite the same as for a TMA; you will find specific advice on exam essays in Unit 16.

Outline of the process

There are three stages:

1. Analyse the question.
2. Draw up a plan.
3. Produce final version.

1   Analysing the question

Make sure you understand, not only *what* you are being asked to write about, but *how* you are being asked to present your material. An essay question is not simply a licence to 'write all you know about X'. Note carefully, for instance, whether the question requires you to 'compare and contrast' (that is, bring out the similarities and differences between), 'demonstrate' (that is, prove by example), 'discuss' (that is, describe and evaluate the pros and cons), 'state' (that is, set out without explanation or background), 'explain', 'outline', 'summarize' or 'illustrate' (this can be with examples or diagrams or both, depending on the context.)

2   Drawing up a plan

However pressed for time, do not skip this. Decide on the main points that will form the framework of your essay, then on the subsidiary or illustrative points. Some people like to lay out plans in the form of headings and subheadings; others prefer them to resemble flow-charts, with boxes and arrows showing the connections between ideas. Once you have drawn up the basic plan, go through it adding numbers to indicate the order in which the topics will be presented. Make sure this involves a logical progression from one point to the next.

3   Producing the final version

Don't forget to include an introductory paragraph. This needn't be long — just a few sentences to explain the topic and the general direction in which you are about to take the reader.

Likewise, don't let the essay tail off at the end without a proper conclusion. Again, a very short summary paragraph is all that is required.

The essay should form a coherent whole. Make sure that there is a logical progression of ideas, and that each section is linked to the preceding one.

It is perfectly acceptable to use section headings if these help you and the reader to keep track of the arguments. Remember that marks will be specifically allocated to structure and presentation: you will forfeit these if you allow the essay to ramble.

Each topic must be covered at a length and depth appropriate to its importance in the overall argument. Make sure that the crucial areas receive the most detailed coverage and are supported by a relevant example. Topics of similar importance should be allocated roughly equal amounts of space.

As a last checklist, the following may be helpful:

(a) Do include diagrams, tables and equations when relevant, but remember to define all the symbols you use.

(b) Use headings when appropriate; otherwise, break the material up into manageable paragraphs for the reader.